OBSERVATIONS ON THE GASTRIC JUICE

EXPERIMENTS AND OBSERVATIONS

ON THE

GASTRIC JUICE AND THE PHYSIOLOGY OF DIGESTION

BY

WILLIAM BEAUMONT, M.D.
SURGEON IN THE UNITED STATES ARMY

FACSIMILE OF THE ORIGINAL EDITION OF 1833
TOGETHER WITH A BIOGRAPHICAL ESSAY

A PIONEER AMERICAN PHYSIOLOGIST

BY SIR WILLIAM OSLER

DOVER PUBLICATIONS, INC.
NEW YORK

This new Dover edition first published in 1959 is an unabridged and unaltered republication of the original edition (1833) together with a reprint of Dr. Osler's address on Beaumont delivered before the St. Louis Medical Society on October 4, 1902. Dover Publications is grateful to Dr. John F. Fulton for kindly loaning his copy of the original edition for this purpose.

Manufactured in the United States of America
Dover Publications, Inc.
180 Varick Street
New York 14, New York

WILLIAM BEAUMONT

BY

WILLIAM OSLER, M.D.

WILLIAM BEAUMONT.

A PIONEER AMERICAN PHYSIOLOGIST.*

Come with me for a few moments on a lovely June day in 1822, to what were then far-off northern wilds, to the Island of Michilimacinac, where the waters of Lake Michigan and Lake Huron unite and where stands Fort Mackinac, rich in the memories of Indian and voyageur, one of the four important posts on the upper lakes in the days when the rose and the fleur-de-lys strove for the mastery of the western world. Here the noble Marquette labored for his Lord, and here beneath the chapel of St. Ignace they laid his bones to rest. Here the intrepid LaSalle, the brave Tonty and the resolute Du Luht had halted in their wild wanderings. Its palisades and block-houses had echoed the war-whoops of Ojibwas and Ottawas, of Hurons and Iroquois, and the old fort had been the scene of bloody massacres and hard-fought fights, but at the conclusion of the War of 1812, after two centuries of struggle, peace settled at last on the island. The fort was occupied by United States troops, who kept the Indians in check and did general police duty on the frontier, and the place had become a rendezvous for Indians and voyageurs in the employ of the American Fur Company. On this bright spring morning the village presented an animated scene. The annual return tide to the trading post was in full course, and the beach was thronged with canoes and batteaux

* An Address before the St. Louis Medical Society, Oct. 4, 1902.

laden with the pelts of the winter's hunt. Voyageurs and Indians, men, women and children, with here and there a few soldiers, made up a motley crowd. Suddenly from the company's store there is a loud report of a gun, and amid the confusion and excitement the rumor spreads of an accident, and there is a hurrying of messengers to the barracks for a doctor. In a few minutes (Beaumont says twenty-five or thirty, an eye-witness says three) an alert-looking man in the uniform of a U. S. Army surgeon made his way through the crowd and was at the side of a young French Canadian who had been wounded by the discharge of a gun, and with a composure bred of an exceptional experience of such injuries, prepared to make the examination. Though youthful in appearance, Surgeon Beaumont had seen much service, and at the capture of York and at the investment of Plattsburgh he had shown a coolness and bravery under fire which had won high praise from his superior officers. The man and the opportunity had met — the outcome is my story of this evening.

I. THE OPPORTUNITY — ALEXIS ST. MARTIN.

On the morning of June 6 a young French Canadian, Alexis St. Martin, was standing in the company's store, "where one of the party was holding a shotgun (not a musket), which was accidentally discharged, the whole charge entering St. Martin's body. The muzzle was not over three feet from him — I think not more than two. The wadding entered, as well as pieces of his clothing; his shirt took fire; he fell, as we supposed, dead."

"Doctor Beaumont, the surgeon of the fort, was immediately sent for and reached the wounded man in a very short time, probably three minutes. We had just gotten him on a cot and were taking off some of his clothing. After the doctor had extracted part of the shot, together with pieces of clothing, and dressed his wound carefully, Robert Stuart and others assisting, he left him, remarking: 'The man can not live thirty-six hours; I will come and see him by and by.' In two or three hours he visited him again, expressing surprise at finding him doing better than he had anticipated. The next day, after getting out more shot and clothing and cutting off ragged edges of the wound, he informed Mr. Stuart, in my presence, that he thought he would recover." *

The description of the wound has been so often quoted as reported in Beaumont's work that I give here the interesting summary which I find in a "Memorial" presented to the Senate and House of Representatives by Beaumont. "The wound was received just under the left breast, and supposed, at the time, to have been mortal. A large portion of the side was blown off, the ribs fractured and openings made into the cavities of the chest and abdomen, through which protruded portions of the lungs and stomach, much lacerated and burnt, exhibiting altogether an appalling and hopeless case. The diaphragm was lacerated and a perforation made directly into the cavity of the

* Statement of G. G. Hubbard, an officer of the company, who was present when St. Martin was shot, quoted by Dr. J. R. Baily, of Mackinac Island, in his address on the occasion of the Beaumont Memorial Exercises, Mackinac Island, July 10, 1900. The Physician and Surgeon, December, 1900.

stomach, through which food was escaping at the time your memorialist was called to his relief. His life was at first wholly despaired of, but he very unexpectedly survived the immediate effects of the wound, and necessarily continued a long time under the constant professional care and treatment of your memorialist, and, by the blessing of God, finally recovered his health and strength.

"At the end of about ten months the wound was partially healed, but he was still an object altogether miserable and helpless. In this situation he was declared 'a common pauper' by the civil authorities of the county, and it was resolved by them that they were not able, nor required, to provide for or support, and finally declined taking care of him, and, in pursuance of what they probably believed to be their public duty, authorized by the laws of the territory, were about to transport him, in this condition, to the place of his nativity in lower Canada, a distance of more than fifteen hundred miles.

"Believing the life of St. Martin must inevitably be sacrificed if such attempt to remove him should be carried into execution at that time, your memorialist, after earnest, repeated, but unavailing,remonstrances against such a course of proceedings, resolved, as the only way to rescue St. Martin from impending misery and death, to arrest the process of transportation and prevent the consequent suffering, by taking him into his own private family, where all the care and attention were bestowed that his condition required.

"St. Martin was, at this time, as before intimated, altogether helpless and suffering under the debilitating effects of his wounds — naked and destitute of everything. In this situation your memorialist received, kept, nursed, medically and surgically treated and sustained him, at much inconvenience and expense, for nearly two years, dressing his wounds daily, and for considerable part of the time twice a day, nursed him, fed him, clothed him, lodged him and furnished him with such necessaries and comforts as his condition and suffering required.

"At the end of these two years he had become able to walk and help himself a little, though unable to provide for his own necessities. In this situation your memorialist retained St. Martin in his family for the special purpose of making physiological experiments."

In the month of May, 1825, Beaumont began the experiments. In June he was ordered to Fort Niagara, where, taking the man with him, he continued the experiments until August. He then took him to Burlington and to Plattsburgh. From the latter place St. Martin returned to Canada, without obtaining Dr. Beaumont's consent. He remained in Canada four years, worked as a voyageur, married and had two children. In 1829 Beaumont succeeded in getting track of St. Martin, and the American Fur Company engaged him and transported him to Fort Crawford on the upper Mississippi. The side and wound were in the same condition as in 1825. Experiments were continued uninterruptedly until March, 1831, when circumstances made it expedient that he should return with his family to lower Canada. The "circumstances," as we gather from letters, were the discontent and homesickness of his wife. As illustrating the mode of travel, Beaumont states that St. Martin took his family in an open canoe "via the Mississippi, passing by St. Louis, ascended the Ohio river, then crossed the state of Ohio to the lakes, and descended the Erie and Ontario and the river St. Lawrence to Montreal, where they arrived in June." Dr. Beaumont often lays stress on the physical vigor of St. Martin as showing how completely he had recovered from the wound. In November, 1832, he again engaged himself to submit to another series of experiments in Plattsburgh and Washington. The last recorded experiment is in November, 1833.

Among the Beaumont papers, for an examination of which I am much indebted to his daughter, Mrs. Keim (Appendix A), there is a large mass of correspondence relating to St. Martin, extending from 1827, two years after he had left the doctor's employ, to October, 1852. Alexis was in Dr. Beaumont's employ in the periods already specified. In 1833 he was enrolled in the United States Army at Washington as Sergeant Alexis St. Martin, of a detachment of orderlies stationed at the War Department. He was then 28 years of age, and was five feet five inches in height.

Among the papers there are two articles of agreement, both signed by the contracting parties, one dated Oct. 19, 1833, and the other November 7 of the same year. In the former he bound himself for a term of one year to "serve, abide and continue with the said William Beaumont, wherever he shall go or travel or reside in any part of the world his covenant servant and diligently and faithfully, etc., . . . that he, the said Alexis, will at all times during said term when thereto directed or required by said William, submit to assist and promote by all means in his power such philosophical or medical experiments as the said William shall direct or cause to be made on or in the stomach of him, the said Alexis, either through and by means of the aperture or opening thereto in the side of him, the said Alexis, or otherwise, and will obey, suffer and comply with all reasonable and proper orders of or experiments of the said William in relation thereto and in relation to the exhibiting and showing of his said stomach and the powers and properties thereto and of the appurtenances and the powers, properties and situation and state of the contents thereof." The agreement was that he should be paid his board and lodging and $150 for the year. In the other agreement it is for two years and the remuneration $400. He was paid a certain amount of the money down.

There are some letters from Alexis himself, all written for him and signed with his mark. In June, 1834, he writes that his wife was not willing to let him go and thinks that he can do a great deal better to stay at home. From this time on Alexis was never again in Dr. Beaumont's employ.

There is a most interesting and protracted correspondence in the years 1836, 1837, 1838, 1839, 1840, 1842, 1846, 1851 and 1852, all relating to attempts to induce Alexis to come to St. Louis. For the greater part of this time he was in Berthier, in the district of Montreal, and the correspondence was chiefly conducted with a Mr. William Morrison, who had been in the northwest fur trade and who took the greatest interest in Alexis and tried to induce him to go to St. Louis. (See Appendix B.)

In 1846 Beaumont sent his son Israel for Alexis, and in a letter dated Aug. 9, 1846, his son writes from Troy: "I have just returned from Montreal, but without Alexis. Upon arriving at Berthier I found that he owned and lived on a farm about fifteen miles sou.-west of the village." Nothing would induce him to go.

The correspondence with Mr. Morrison in 1851 and 1852 is most voluminous, and Dr. Beaumont offered Alexis $500 for the year, with comfortable support for his family. He agreed at one time to go, but it was too late in the winter and he could not get away.

The last letter of the series is dated Oct. 15, 1852, and is from Dr. Beaumont to Alexis, whom he addresses as *Mon Ami*. Two sentences in this are worth quoting: "Without reference to past efforts and disappointments — or expectation of ever obtaining your services again for the purpose of experiments, etc., upon the proposals and conditions heretofore made and suggested, I now proffer to you in faith and sincerity, new, and I hope satisfactory, terms and conditions to ensure your prompt and faithful compliance with my most fervent desire to have you again with me — not only for my own individual gratification, and the benefits of medical science, but also for your own and family's present good and future welfare." He concludes with, "I can say no more, Alexis — you know what I *have* done for you many years since — what I have been *trying*, and am still anxious and wishing to do with and for you — what efforts, anxieties, anticipations and disappointments I have suffered from your non-fulfilment of my expectations. Don't disappoint me more nor forfeit the bounties and blessings reserved for you."

So much interest was excited by the report of the experiments that it was suggested to Beaumont that he should take Alexis to Europe and submit him there to a more extended series of observations by skilled physiologists. Writing June 10, 1833, he says: "I shall engage him for five or six years if he will agree, of which I expect there is no doubt. He has always been pleased with the idea of going to France. I feel much gratified at the expression of Mr. Livingston's desire that we should visit Paris, and shall duly consider the interest he takes in the subject and make the best arrangements I can to meet his views and yours." Mr. Livingston, the American minister, wrote from Paris March 18, 1834, saying that he had submitted the work to Orfila and the Academy of Sciences, which had appointed a committee to determine if additional experiments were necessary and whether it was advisable to send to America for Alexis. Nothing, I believe, ever came of this, nor, so far as I can find, did Alexis visit Paris. Other attempts were made to secure him for purposes of study. In 1840 a student of Dr. Beaumont's, George Johnson, then at the University of Pennsylvania, wrote saying that Dr. Jackson had told him of efforts made to get Alexis to London, and Dr. Gibson informed him that the Medical Society of London had raised £300 or £400 to induce St. Martin to come, and that he, Dr. Gibson, had been trying to find St. Martin for his London friends. There are letters in the same year from Dr. R. D. Thomson of London to Professor Silliman urging him to arrange that Dr. Beaumont and Alexis should visit London. In 1856 St. Martin was under the observation of Dr. Francis Gurney Smith, in Philadelphia, who reported a brief series of experiments, so far as I know the only other report made on him.*

St. Martin had to stand a good deal of chaffing about the hole in his side. His comrades called him "the man with a lid on his stomach." In his memorial address Mr. C. S. Osborn of Sault Ste. Marie states that Miss Catherwood tells a story of Etienne St. Martin fighting with Charlie Charette because Charlie ridiculed his brother. Etienne stabbed him severely and swore that he would kill the whole brigade if they did not stop deriding his brother's stomach.

At one time St. Martin traveled about exhibiting the wound to physicians, medical students and before medical societies. In a copy of Beaumont's work, formerly belonging to Austin Flint, Jr., and now in the possession of a physician of St. Louis, there is a photograph of Alexis sent to Dr. Flint. There are statements made that he went to Europe, but of such a visit I can find no record.

My interest in St. Martin was of quite the general character of a teacher of physiology, who every session referred to his remarkable wound and showed

* Medical Examiner, 1856, and Experiments on Digestion, Phila., 1856.

Beaumont's book with the illustration. In the spring of 1880, while still a resident of Montreal, I saw a notice in the newspapers of his death at St. Thomas. I immediately wrote to a physician and to the parish priest, urging them to secure me the privilege of an autopsy and offering to pay a fair sum for the stomach, which I agreed to place in the Army Medical Museum in Washington, but without avail. Subsequently, through the kindness of the Hon. Mr. Justice Baby, I obtained the following details of St. Martin's later life, and the picture here given, which was taken the year before his death so as to show the wound, which I here show you. Judge Baby writes to his friend, Prof. D. C. MacCallum of Montreal, as follows: "I have much pleasure to-day in placing in your hands such information about St. Martin as Revd. Mr. Chicoine, Curé of St. Thomas, has just handed over to me. Alexis Bidigan, *dit* St. Martin, died at St. Thomas de Joliette on the 24th of June, 1880, and was buried in the cemetery of the parish on the 28th of the same month. The last sacraments of the Catholic Church were ministered to him by the Revd. Curé Chicoine, who also attended at his burial service. The body was then in such an advanced stage of decomposition that it could not be admitted into the church, but had to be left outside during the funeral service. The family resisted all requests — most pressing as they were — on the part of the members of the medical profession for an autopsy, and also kept the body at home much longer than usual and during a hot spell of weather, so as to allow decomposition to set in and baffle, as they thought, the doctors of the surrounding country and others. They had also the grave dug eight feet below the surface of the ground in order to prevent any attempt at a resurrection. When he died St. Martin was 83 years of age, and left a widow, whose maiden name was Marie Joly. She survived him by nearly seven years, dying at St. Thomas on the 20th of April, 1887, at the very old age of 90 years. They left four children still alive — Alexis, Charles, Henriette and Marie.

"Now I may add the following details for myself. When I came to know St. Martin it must have been a few years before his death. A law suit brought him to my office here in Joliette. I was seized with his interests; he came to my office a good many times, during which visits he spoke to me at great length of his former life, how his wound had been caused, his peregrinations through Europe and the United States, etc. He showed me his wound. He complained bitterly of some doctors who had awfully misused him, and had kind words for others. He had made considerable money during his tours, but had expended and thrown it all away in a frolicsome way, especially in the old country. When I came across him he was rather poor, living on a small, scanty farm in St. Thomas, and very much addicted to drink, almost a drunkard one might say. He was a tall, lean man, with a very dark complexion, and appeared to me then of a morose disposition."

II. THE BOOK.

In the four periods in which Alexis had been under the care and study of Beaumont a large series of observations had been recorded, amounting in all to 238. A preliminary account of the case and of the first group of observations appeared in the *Philadelphia Medical Recorder* in January, 1825. During the stay in Washington in 1832 the great importance of the observations had become impressed on the Surgeon-General, Dr. Lovell, who seems to have acted in a most generous and kindly spirit. Beaumont tried to induce him to undertake the arrangement of the observations, but Lovell insisted that he should do the work himself. In the spring of 1833 Alexis was taken to New York and there shown to the prominent members of the profession, and careful drawings and colored sketches were made of the wound by Mr. King. A prospectus of the work was issued and was distributed by the Surgeon-General, who speaks in a letter of sending them to Dr. Franklin Bache and to Dr. Stewart of Philadelphia, and in a letter from Dr. Bache to Dr. Beaumont acknowledging the receipt of a bottle of gastric juice, Bache states that he has placed the prospectus in Mr. Judah Dobson's store and has asked for subscribers. Beaumont did not find New York a very congenial place. He complained of the difficulty of doing the work owing to the vexatious social intercourse. He applied for permission to go to Plattsburgh, in order to complete the book. After having made inquiries in New York and Philadelphia about terms of publication he decided, as the had to be issued at his own expense, that it could be as well and much more cheaply printed at Plattsburg, where he would also have the advice and help of his cousin, Dr. Samuel Beaumont. In a letter to the Surgeon-General, dated June 10, 1833, he acknowledges the permission to go to Plattsburgh, and says: "I shall make my arrangements to leave here for Pl. in about a week to *rush* the execution of the Book as fast as possible. I am now having the drawings taken by Mr. King engraved here."

The summer was occupied in making a fresh series of experiments and getting the work in type. On December 3 he writes the Surgeon-General that the book will be ready for distribution in a few days and that 1,000 copies will be printed.

The work is an octavo volume of 280 pages, entitled "Experiments and Observations on the Gastric Juice and the Physiology of Digestion," by William Beaumont, M.D., Surgeon in the United States Army. Plattsburgh. Printed by F. P. Allen, 1833. While it is well and carefully printed, the paper and type are not of the best, and one can not but regret that Beaumont did not take the advice of Dr. Franklin Bache, who urged him strongly not to have the work printed at Plattsburgh, but in Philadelphia, where it could be done in very much better style. The dedication of the work to Joseph Lovell, M.D., Surgeon-General of the United States Army, acknowledges in somewhat laudatory terms the debt which Beaumont felt he owed to his chief, who very gratefully acknowledges the compliment and the kindly feeling, but characterizes the dedication as "somewhat apocryphal."

The work is divided into two main portions; first, the preliminary observations on the general physiology of digestion in seven sections: Section I, Of Aliment; Section II, Of Hunger and Thirst; Section III, Of Satisfaction and Satiety; Section IV, Of Mastication, Insalivation and Deglutition; Section V, Of Digestion by the Gastric Juice; Section VI, Of the Appearance of the Villous Coat, and of the Motions of the Stomach; Section VII, Of Chylification and Uses of the Bile and Pancreatic Juice. The greater part of the book is occupied by the larger section of the detailed account of the four series of experiments and observations. The work concludes with a series of 51 inferences from the foregoing experiments and observations.

The subsequent history of the book itself is of interest, and may be dealt with here. In 1834 copies of the Plattsburgh edition, printed by F. P. Allen, were issued by Lilly, Wait & Co., of Boston.

In the Beaumont correspondence there are many letters from a Dr. McCall, in Utica, N. Y., who was an intimate friend of a Mr. Wm. Combe, a brother of the well-known physiologist and popular writer, Dr. Andrew Combe of Edinburgh. Doubtless it was through this connection that in 1838 Dr. Combe issued an edition in Scotland, with numerous notes and comments. (Appendix C.)

The second edition was issued from Burlington, Vt., in 1847, with the same title page, but after Second Edition there are the words, Corrected by Samuel Beaumont, M.D., who was Dr. William Beaumont's cousin. In the preface to this edition the statement is made that the first edition, though a large one of 3,000 copies, had been exhausted. This does not agree with the statement made in a letter of Dec. 3, 1833, to the Surgeon-General, stating that the edition was to be 1,000 copies. Of course more may have been printed before the type was distributed. While it is stated to be a new and improved edition, so far as I can gather it is a verbatim reprint, with no additional observations, but with a good many minor corrections. In an appendix (D) I give an interesting letter from Dr. Samuel Beaumont with reference to the issue of this edition.

A German edition was issued in 1834 with the following title: "Neue Versuche und Beobachtungen ueber den Magensaft und die Physiologie der Verdauung, auf eine hochst merkwurdige Weise wahrend einer Reihe von 7 Jahren, an einen und demselben Subject angestellt." Beaumont's earlier paper, already referred to, was abstracted in the Magazin der auslandischen Litteratur der gesammten Heilkunde, Hamburg, 1826, and also in the Archives generales de Medecine, Paris, 1828. I can not find that there was a French edition of the work.

The "Experiments and Observations" attracted universal attention, both at home and abroad. The journals of the period contained very full accounts of the work, and within a few years the valuable additions to our knowledge filtered into the text-books of physiology, in which to-day in certain descriptions of the gastric juice and of the phenomena of digestion even the very language of the work is copied.

III. THE VALUE OF BEAUMONT'S OBSERVATIONS.

There had been other instances of artificial gastric fistula in man which had been made the subject of experimental study, but the case of St. Martin stands out from all others on account of the ability and care with which the experiments were conducted. As Dr. Combe says, the value of these experiments consists partly in the admirable opportunities for observation which Beaumont enjoyed, and partly in the candid and truth-seeking spirit in which all his inquiries seem to have been conducted. "It would be difficult to point out any observer who excels him in devotion to truth and freedom from the trammels of theory or prejudice. He tells plainly what he saw and leaves every one to draw his own inferences, or where he lays down conclusions he does so with a degree of modesty and fairness of which few perhaps in his circumstances would have been capable."

To appreciate the value of Beaumont's studies it is necessary to refer for a few moments to our knowledge of the physiology of digestion in the year 1832, the date of the publication. Take, for example, "The Work on Human Physiology" (published in the very year of the appearance of Beaumont's book), by Dunglison, a man of wide learning and thoroughly informed in the literature of the subject. The five or six old theories of stomach digestion, concoction, putrefaction, trituration, fermentation and maceration, are all discussed, and Wm. Hunter's pithy remark is quoted, "some physiologists will have it, that the stomach is a mill, others, that it is a fermenting vat, others, again, that it is a stew-pan; but, in my view of the matter, it is neither a mill, a fermenting vat nor a stew-pan; but a stomach, gentlemen, a stomach."

The theory of chemical solution is accepted. This had been placed on a sound basis by the experiments of Reaumur, Spallanzani and Stevens, while the studies of Tiedemann and Gmelin and of Prout had done much to solve the problems of the chemistry of the juice. But very much uncertainty existed as to the phenomena occurring during digestion in the stomach, the precise mode of action of the juice, the nature of the juice itself and its action outside the body. On all these points the observations of Beaumont brought clearness and light where there had been previously the greatest obscurity.

The following may be regarded as the most important of the results of Beaumont's observations: First, the accuracy and completeness of description of the gastric juice itself. You will all recognize the following quotation, which has entered into the textbooks and passes current to-day. "Pure gastric juice, when taken directly out of the stomach of a healthy adult, unmixed with any other fluid, save a portion of the mucus of the stomach with which it is most commonly and perhaps always combined, is a clear, transparent fluid; inodorous; a little saltish, and very perceptibly acid. Its taste, when applied to the tongue, is similar to this mucilaginous water slightly acidulated with muriatic acid. It is readily diffusible in water, wine or spirits; slightly effervesces with alkalies; and is an effectual solvent of the *materia alimentaria*. It possesses the property of coagulating albumen, in an eminent degree; is powerfully antiseptic, checking the putrefaction of meat; and effectually restorative of healthy action, when applied to old, foetid sores and foul, ulcerating surfaces."

Secondly, the confirmation of the observation of Prout that the important acid of the gastric juice was the muriatic or hydrochloric. An analysis of St. Martin's gastric juice was made by Dunglison, at that time a professor in the University of Virginia, and by Benjamin Silliman of Yale, both of whom determined the presence of free hydrochloric acid. A specimen was sent to the distinguished Swedish chemist, Berzelius, whose report did not arrive in to time be included in the work. In a letter dated July 19, 1834, he writes to Professor Silliman that he had not been able to make a satisfactory analysis of the juice. The letter is published in *Silliman's Journal*, Vol. 27, July, 1835.

Thirdly, the recognition of the fact that the essential elements of the gastric juice and the mucus were separate secretions.

Fourthly, the establishment by direct observation of the profound influence on the secretion of the gastric juice and on digestion of mental disturbances.

Fifthly, a more accurate and fuller comparative study of the digestion in the stomach with digestion outside the body, confirming in a most elaborate series of experiments the older observations of Spallanzani and Stevens.

Sixthly, the refutation of many erroneous opinions relating to gastric digestion and the establishment of a number of minor points of great importance, such as, for instance, the rapid disappearance of water from the stomach through the pylorus, a piont brought out by recent experiments, but insisted on and amply proven by Beaumont.

Seventhly, the first comprehensive and thorough study of the motions of the stomach, observations on which, indeed, are based the most of our present knowledge.

And lastly, a study of the digestibility of different articles of diet in the stomach, which remains to-day one of the most important contributions ever made to practical dietetics.

The greater rapidity with which solid food is digested, the injurious effects on the stomach of tea and coffee, when taken in excess, the pernicious influence of alcoholic drinks on the digestion, are constantly referred to. An all-important practical point insisted on by Beaumont needs emphatic reiteration to this generation: "The system requires much less than is generally supplied to it. The stomach disposes of a definite quantity. If more be taken than the actual wants of the economy require, the residue remains in the stomach and becomes a source of irritation and produces a consequent aberration of function, or passes into the lower bowel in an undigested state, and extends to them its deleterious influence. Dyspepsia is oftener the effect of over-eating and over-drinking than of any other cause."

One is much impressed, too, in going over the experiments, to note with what modesty Beaumont refers to his own work. He speaks of himself as a humble "enquirer after truth and a simple experimenter." "Honest objections, no doubt, are entertained against the doctrine of digestion by the gastric juice. That they are so entertained by these gentlemen I have no doubt. And I cheerfully concede to them the merit of great ingenuity, talents and learning, in raising objections to the commonly received hypothesis, as well as ability in maintaining their peculiar opinions. But we ought not to allow ourselves to be seduced by the ingenuity of argument or the blandishments of style. Truth, like beauty, when 'unadorned is adorned the most'; and in prosecuting these experiments and inquiries, I believe I have been guided by its light. Facts are more persuasive than arguments, however ingeniously made, and by their eloquence I hope I have been able to plead for the support and maintenance of those doctrines which have had for their advocates such men as Sydenham, Hunter, Spallanzani, Richerand, Abernethy, Broussais, Philip, Paris, Bostock, the Heidelburg and Paris professors, Dunglison, and a host of other luminaries in the science of physiology."

In reality Beaumont anticipated some of the most recent studies in the physiology of digestion. Doubtless many of you have heard of Professor Pawlow's, of St. Petersburg, new work on the subject. It has been translated into German, and I see that an English edition is advertised. He has studied the gastric juice in an isolated pouch, ingeniously made at the fundus of the stomach of the dog, from which the juice could be obtained in a pure state. One of his results is the very first announced by Beaumont and confirmed by scores of observations on St. Martin, viz., that, as he says, "the gastric juice never appears to be accumulated in the cavity of the stomach while fasting." Pawlow has shown very clearly that there is a relation between the amount of food taken and the quantity of gastric juice secreted. Beaumont came to the same conclusion: "when aliment is received the juice is given in exact proportion to its requirements for solution." A third point on which Pawlow lays stress is the curve of secretion of the gastric juice, the manner in which it is poured out during digestion. The greatest secretion, he has shown, takes place in the earlier hours. On this point hear Beaumont: "It (the gastric juice) then begins to exude from the proper vessels and increases in proportion to the quantity of aliment naturally required and received." And again: "When a due and moderate supply of food has been received it is probable that the whole quantity of gastric juice for its complete solution is secreted and mixed with it in a short time." A fourth point, worked out beautifully by Powlow, is the adaptation of the juice to the nature of the food, on which I do not see any reference by Beaumont, but there are no experiments more full than those in which he deals with the influence of exercise, weather and the emotions on the quantity of the juice secreted.

IV. MAN AND DOCTOR.

Sketches of Dr. Beaumont's life have appeared from time to time. There is a worthy memoir by Dr. T. Reyburn in the *St. Louis Medical and Surgical*

Journal, 1854, and Dr. A. J. Steele, at the first annual commencement of the Beaumont Medical College, 1887, told well and graphically the story of his life. A few years ago Dr. Frank J. Lutz, of this city, sketched his life for the memorial meeting of the Michigan State Medical Society on the occasion of the dedication of a Beaumont monument.

Among the papers kindly sent to me by his daughter, Mrs. Keim, are many autobiographical materials, particularly relating to his early studies and to his work as a surgeon in the War of 1812. There is an excellent paper in the handwriting, it is said, of his son, giving a summary of the earlier period of his life. So far as I know this has not been published, and I give it in full:

Dr. William Beaumont was born in the town of Lebanon, Conn., on the 21st day of November, A. D. 1785. His father was a thriving farmer and an active politician of the proud old Jeffersonian school, whose highest boast was his firm support and strict adherence to the honest principles he advocated. William was his third son, who, in the winter of 1806–7, in the 22d year of his age, prompted by a spirit of independence and adventure, left the paternal roof to seek a fortune and a name. His outfit consisted of a horse and cutter, a barrel of cider, and one hundred dollars of hard-earned money. With this he started, laying his course northwardly, without any particular destination, Honor his rule of action, Truth his only landmark, and trust placed implicity in Heaven. Traversing the western part of Massachusetts and Vermont in the spring of 1807 he arrived at the little village of Champlain, N. Y., on the Canada frontier — an utter stranger, friendless and alone. But honesty of purpose and true energy invariably work good results. He soon gained the people's confidence and was entrusted with their village school, which he conducted about three years, devoting his leisure hours to the study of medical works from the library of Dr. Seth Pomeroy, his first patron. He then went over to St. Albans, Vt., where he entered the office of Dr. Benjamin Chandler and commenced a regular course of medical reading, which he followed for two years, gaining the utmost confidence and esteem of his kind preceptor and friends. About this time the War of 1812 commenced, and he applied for an appointment in the U. S. Army, successfully. He was appointed assistant-surgeon to the Sixth Infantry, and joined his regiment at Plattsburgh, N. Y., on the 13th of September, 1812. On the 19th of March, 1813, he marched from Plattsburgh with the First Brigade, for Sackett's Harbor, where they arrived on the 27th inst. Here he remained in camp till the 22d of April, when he embarked with the troops on Lake Ontario. His journal will best tell this portion of his history:

"April 22, 1813. — Embarked with Captain Humphreys, Walworth and Muhlenburg, and companies on board the Schooner 'Julia.' The rest of the brigade, and the Second, with Foresith's Rifle Regiment and the Eighth Artillery, on board a ship, brig and schooner — remain in the harbor till next morning.

"23d.— 11 o'clock a. m.— Weighs anchor and put out under the impression we were going to Kingston. Got out 15 or 20 miles — encountered a storm — wind ahead and the fleet returned to harbor.

"24th.— 6 o'clock a. m.— Put out with a fair wind — mild and pleasant— the fleet sailing in fine order.

"26th.— Wind pretty strong — increasing — waves run high, tossing our vessels roughly. At half past four pass the mouth of Niagara river. This circumstance baffles imagination as to where we are going — first impressed with the idea of Kingston — then to Niagara —but now our destination must be 'Little York.' At sunset came in view of York Town and the Fort, where we lay off some 3 or 4 leagues for the night.

"27th.— Sailed into harbor and came to anchor a little below the British Garrison. Filled the boats and effected a landing, though not without difficulty and the loss of some men. The British marched their troops down the beach to cut us off as landing, and, though they had every advantage, they could not effect their design. A hot engagement ensued, in which the enemy lost nearly a third of their men and were soon compelled to quit the field, leaving their dead and wounded strewn in every direction. They retired to the Garrison, but from the loss sustained in the engagement, the undaunted courage of our men, and the brisk firing from our fleet, with the 12 and 32 pounders, they were soon obliged to evacuate it and retreat with all possible speed.— Driven to this alternative they devised the inhuman project of blowing up their magazine, containing 300 pounds of powder, the explosion of which had well-nigh destroyed our army. Over 300 were wounded and about 60 killed on the spot, by stones of all dimensions talling, like a shower of hail, in the midst of our ranks. A most distressing scene ensues in the hospital. Nothing is heard but the agonizing groans and supplications of the wounded and the dying. The surgeons wade in blood cutting off arms and legs and trepaning heads, while the poor sufferers cry, 'O, my God! Doctor, relieve me from this misery! I can not live!' 'Twas enough to touch the veriest heart of steel and move the most relentless savage. Imagine the shocking scene, where fellow-beings lie mashed and mangled — legs and arms broken and sundered — heads and bodies bruised and mutilated to disfigurement! My deepest sympathies were roused — I cut and slashed for 36 hours without food or sleep.

"29th.— Dressed upwards of 50 patients — from simple contusions to the worst of compound fractures — more than half the latter. Performed two cases of amputation and one of trepaning. At 12 p. m. retired to rest my fatigued body and mind."

One month after the taking of York he witnessed the storming of Fort George. The troops were transported from York to "Four-Mile Creek" (in the vicinity of Ft. George), where they encamped from the 10th of May to the 27th, when they advanced to the attack. His journal runs thus:

"May 27 (1813).— Embarked at break of day — Col. Scott with 800 men, for the advanced guard, supported by the First Brigade, commanded by General Boyd, moved in concert with the shipping to the enemy's shore and landed under their battery and in front of their fire with surprising success, not losing more than 30 men in the engagement, though the enemy's whole force was placed in the most advantageous situation possible. We routed them from their chosen spot — drove them from the country and took possession of the town and garrison."

On the 11th of September, 1814, he was at the Battle of Plattsburgh, still serving as assistant-surgeon, though doing all the duty of a full surgeon. At the close of the war, in 1815, when the Army was cut down, he was retained in service, but resigned soon after, deeming himself unjustly treated by the government in having others, younger and less experienced, promoted over him.

In 1816 he settled in Plattsburgh and remained there four years in successful practice. In the meantime his army friends had persuaded him to join the service again, and, having applied, he was reappointed, in 1820, and ordered to Ft. Mackinac as post surgeon. At the end of the first year he obtained leave of absence, returned to Plattsburgh and married one of the most amiable and interesting ladies of that place. (She still survives her honored husband, and in her green old age is loved devotedly by all who know her.) He returned to Mackinac the same year, and in 1822 came in possession of Alexis St. Martin, the subject of his "Experiments on the Gastric Juice." By the accidental discharge of his gun, while hunting, St. Martin had dangerously wounded himself in the abdomen and came under the treatment of Dr. Beaumont, who healed the wound (in itself a triumph of skill almost unequaled) and in 1825 commenced a series of experiments, the results of which have a world-wide publication. These experiments were continued, with various interruptions, for eight years, during which time he was ordered from post to post — now at Niagara, N. Y., anon at Green Bay, Mich., and finally at Fort Crawford, on the Mississippi. In 1834 he was ordered to St. Louis, where he remained in service till 1839, when he resigned. He then commenced service with the citizens of St. Louis, and from that time till the period of his last illness, enjoyed an extensive and distinguished practice, interrupted only by the base attacks of a few disgraceful and malicious knaves (self-deemed members of the medical profession) who sought to destroy a reputation which they could not share. They gained nothing except some little unenviable notoriety and they have skulked away like famished wolves, to die in their hiding places.

The dates of Beaumont's commissions in the army are as follows: Surgeon's Mate, Sixth Regiment of Infantry, Dec. 2, 1812; Cavalry, March 27, 1819; Post Surgeon, Dec. 4, 1819; Surgeon First Regiment and Surgeon, Nov. 6, 1826.

From the biographical sketches of Reyburn, Steel and Lutz, and from the personal reminiscences of his friends, Drs. J. B. Johnson, S. Pollak and Wm. McPheeters, who fortunately remains with you, full of years and honors, we gather a clearly-defined picture of the latter years of his life. It is that of a faithful, honest, hard-working practitioner, doing his duty to his patients, and working with zeal and ability for the best interests of the profession. The strong common sense which he exhibited in his experimental work made him a good physician and a trusty adviser in cases of surgery. Among his letters there are some interesting pictures of his life, particularly in his letters to his cousin, Dr. Samuel Beaumont. Writing to him April 4, 1846, he says:

I have a laborious, lucrative and increasing practice, more than I can possibly attend to, though I have an assistant, Dr. Johnson, a young man who was a pupil of mine from 1835 to 1840. He then went to Philadelphia a year or two to attend lectures, and graduated, and returned here again in 1842, and has been very busy ever since and is so now, but notwithstanding I decline more practice daily than half the doctors in the city get in a week. You thought when you were here before that there was too much competition for you ever to think of succeeding in business here — there is ten times as much now and the better I succeed and prosper for it. You must come with a different feeling from your former — with a determination to follow in my wake and stem the current that I will break for you. I am now in the grand climacteric of life, three-score years and over, with equal or more zeal and ability to do good and contribute to professional service than at forty-five, and I now look forward with pleasing anticipation of success and greater usefulness — have ample competence for ourselves and children, and no doleful or dreaded aspect of the future — to be sure I have to wrestle with some adverse circumstances of life, and more particularly to defend myself against the envious, mean and professional jealousies and the consequent prejudices of some men, but I triumph over them all and go ahead in defiance of them.*

His professional work increased enormously with the rapid growth of the city, but he felt, even in his old age, that delicious exhilaration which it is your pleasure and privilege to enjoy here in the west in a degree rarely experienced by your eastern confrères. Here is a cheery paragraph from a letter dated Oct. 20, 1852: "Domestic affairs are easy, peaceable and pleasant. Health of community good — no severe epidemic diseases prevalent — weather remarkably pleasant — business of all kinds increasing — product of the earth abundant — money plenty — railroads progressing with almost telegraphic speed — I expect to come to Plattsburgh next summer all the way by rail."

But work was becoming more burdensome to a man nearing threescore years and ten, and he expresses it

* He had evidently hopes that when his cousin and son arrived with Alexis they would arrange and plan for another series of experiments and in another year or two make another book, better than the old one.

in another letter when he says: "There is an immense professional practice in this city. I get tired of it, and have been trying hard to withdraw from it altogether, but the more I try the tighter I seem to be held to it by the people. I am actually persecuted, worried and almost worn out with valetudinarian importunities and hypochondriacal groans, repinings and lamentations — Amen."

He continued at work until March, 1853, when he had an accident — a fall while descending some steps. A few weeks later a carbuncle appeared on the neck, and proved fatal April 25. One who knew him well wrote the following estimate (quoted by Dr. F. J. Lutz in his sketch of Beaumont):

"He was gifted with strong natural powers, which, working upon an extensive experience in life, resulted in a species of natural sagacity, which, as I suppose, was something peculiar in him, and not to be attained by any course of study. His temperament was ardent, but never got the better of his instructed and disciplined judgment, and whenever and however employed, he ever adopted the most judicious means for attaining ends that were always honorable. In the sick room, he was a model of patience and kindness, his intuitive perceptions, guiding a pure benevolence, never failed to inspire confidence, and thus he belonged to that class of physicians whose very presence affords Nature a sensible relief."

You do well, citizens of St. Louis and members of our profession, to cherish the memory of William Beaumont. Alive you honored and rewarded him, and there is no reproach against you of neglected merit and talents unrecognized. The profession of the northern part of the state of Michigan has honored itself in erecting a monument to his memory near the scene of his disinterested labors in the cause of humanity and science. His name is linked with one of your educational institutions, and joined with that of a distinguished laborer in another field of practice. But he has a far higher honor than any you can give him here —the honor that can only come when the man and the opportunity meet — and match. Beaumont is the pioneer physiologist of this country, the first to make an important and enduring contribution to this science. His work remains a model of patient, persevering investigation, experiment and research, and the highest praise we can give him is to say that he lived up to and fulfilled the ideals with which he set out and which he expressed when he said: "Truth, like beauty, when 'unadorned, is adorned the most,' and, in prosecuting these experiments and enquiries, I believe I have been guided by its light."

APPENDIX A.

The Beaumont papers in the possession of his daughter, Mrs. Keim of St. Louis, consist of (1) interesting certificates from his preceptors, Dr. Pomeroy and Dr. Chandler, the license from the Third Medical Society of Vermont, the commissions in the U. S. Army, several certificates of honorary membership in societies, and the parchment of the M.D. degree conferred upon him, *honoris causa*, by the Columbian University of Washington, 1833; (2) a journal containing his experiences in the War of 1812, from which I give an extract, a journal of his trip to Fort Mackinac, a journal containing the reports of many cases, among them that of St. Martin (in addition there is a protocol of the case in loose folio sheets), a journal of the experiments, and a commonplace book of receipts and jottings; (3) an extensive correspondence relating to St. Martin and the book, and many rough drafts of sections of the book; (4) a large mass of personal correspondence, much of it of interest as relating to conditions of practice in St. Louis.

The picture reproduced here in his army uniform is from a miniature; the picture which has been previously reproduced is of an older man from a daguerreotype. It is satisfactory to know that the ultimate destination of this most valuable collection of papers is the Surgeon-General's Library of the United States Army, of which Dr. Beaumont was so distinguished an ornament.

APPENDIX B.

On Oct. 20, 1853, he writes to his cousin, Dr. Samuel Beaumont, on the subject of "that old, fistulous Alexis," as he calls him. "Alexis' answer to yours is the very facsimile or stereotype of all his Jesuitical letters to me for the last fifteen years. His object seems only to be to get a heavy bonus and undue advance from me and then disappoint and deceive me, or to palm and impose himself and whole family upon me for support for life.

"I have evaded this design so far; but I verily fear that the strong and increasing impulse of conscious conviction of the great benefits and important usefulness of further and more accurate physiological investigation of the subject will compel me to still further efforts and sacrifices to obtain him. Physiological authors and most able writers on dietetics and gastric functions generally demand it of me in trumpet tones.

"I must have him at all hazards, and obtain the necessary assistance to my individual and private efforts or transfer him to some competent scientific institution for thorough investigation and report — I must retrieve my past ignorance, imbecility and professional remissness of a quarter of a century, or more, by double diligence, intense study and untiring application of soul and body to the subject before I die —

> Should posthumous Time retain my name,
> Let historic truths declare my fame.

"Simultaneous with this I write to Mr. Morrison and Alexis my last and final letters — perhaps, proposing to *him*, as bribe to his cupidity, to give him $500 to come to me *without* his family, for one year — $300 of them for his salary, and $200 for the support and contentment of his family to remain in Canada in the meantime — with the privilege of bringing them on here another year, upon my former proposition of $300 a year, at his own expense and responsibility and support them himself after they get here out of his $300 salary — I think he will take the bait and come on this fall, and when I get him alone again into my keeping and engagement, I will take good care to control him as I please."

APPENDIX C.

Letter from Dr. Andrew Combe, May 1, 1838:

"My Dear Sir — May I beg your acceptance of the accompanying volumes as a small expression of my respect for your character and scientific labors. I need not detain you by repeating in this note the high estimation in which I hold you. The volumes herewith sent will, I trust, convince you of the fact, and that it will not be my fault if you do not receive the credit justly due to your valuable and disinterested services. I remain, My Dear Sir,

Very respectfully yours,

"ANDW. COMBE."

APPENDIX D.

Letter from Dr. Samuel Beaumont, March 16, 1846:

"Your letter of the 1st of February arrived here in the course of mail, and I have attended to the business which you authorized me to do. I am afraid, however, that you will be disappointed, and perhaps dissatisfied with the arrangement. Mr. Goodrich came here some five or six days after I received your letter, and made his proposal, which was to give you every tenth copy for the privilege of publishing an edition. The number he proposed to publish was fifteen hundred, which would give you 150 copies. I did not like to close the bargain on this condition, and he was not disposed to give any more. This was in the evening. I told him to give me time till the next morning, and I would make up my mind. In the morning, after consultation, I concluded to offer him the copyright for the unexpired time (only one year) for two hundred copies. After some demurring, we closed the bargain. I then thought and I still think it was not enough; but it was all I could get. In making up my mind the following considerations presented themselves: "First, that the copyright would expire in one year, and he would then have the right to print it without consulting the author; second, that it would be somewhat mortifying to the author not to have his work republished, even if no great pecuniary benefit was to be obtained by such a republication; and it appeared to me to be quite certain that a new edition would not be soon printed, if I let this opportunity slip; third, I have been long anxious, as I presume you have been, to see the work gotten up in a better dress than it originally had, and in a way which will give it a general credit and more notoriety among all classes of the reading public than it has heretofore possessed — in fact, make it a standard work; fourth, it has given us a chance to give it a thorough correction, a thing which was very desirable. The work, you recollect, was got up in a great hurry, and a great many errors escaped our notice. You may also recollect that the Philadelphia reviewer spoke of the inaccuracies in the work. And he had reason enough for it. In looking over the work critically with a view of correction, I have been perfectly astonished at the errors that occur on almost every page. And although we understood perfectly what we meant to say, the reader would find it somewhat difficult to decipher our meaning. In the first 140 pages I made nearly 300 corrections. These are practically merely verbal alterations or change of phrases or sentences so as to make them more accurate or perspicuous. I have in no case so changed the text as to give it a different meaning. I flatter myself that it will now be more worthy the public patronage; and if for no other, this chance for correction I consider alone almost a sufficient remuneration for the brief limits of the copyright. I have also written a preface for the second edition, making quotations from American and European authorities in praise of the merits of the work. From delicacy I have written this as from the publisher. I think it is pretty well done. The work will probably be published in the course of about a month, and those designed for you will be delivered to me, when I shall send them to you. He guarantees not to sell in the state of Missouri, or the states south and west of that state. But that, of course, is all gammon. The book will be thrown into market, and he can not control the direction in which it will go."

EXPERIMENTS

AND

OBSERVATIONS

ON THE

GASTRIC JUICE,

AND THE

PHYSIOLOGY OF DIGESTION.

BY WILLIAM BEAUMONT, M. D.

Surgeon in the U. S. Army.

PLATTSBURGH,

PRINTED BY F. P. ALLEN.

1833.

TO JOSEPH LOVELL, M. D.

SURGEON GENERAL OF THE UNITED STATES' ARMY,

Whose merit justly entitles him to the rank
which he holds,
And whose zeal in the cause of Medical Science is
equalled only by his ability to promote it.
As a tribute of respect for his public and pri-
vate virtues,
And as a feeble acknowledgement for a long
tried and unvarying friendship,
This work is respectfully dedicated, by
THE AUTHOR.

PREFACE.

THE present age is prolific of works on physiology; therefore in offering to the public another book relative to an important branch of this science, it will perhaps be necessary to assign my motives.

They are, first, a wish to comply with the repeated and urgent solicitations of many medical men who have become partially acquainted with the facts and observations it is my intention to detail; men, in whose judgment I place confidence, and who have expressed their conviction of the deep importance of the experiments, the result of which I mean herewith to submit to the public: secondly, (and it is that which mainly influences me,) my own firm conviction that medical science will be forwarded by the publication.

I am fully aware of the importance of the subject which these experiments are intended to illustrate, as well in a pathological as in a physiological point of view; and I am therefore willing to risk the censure or neglect of critics, if I may be permitted to cast my mite into the treasury of knowledge, and to be the means, either directly or indirectly, of subserving the cause of truth, and ameliorating the condition of suffering humanity.

I make no claim to originality in my opinions, as it respects the existence and operation of the gastric juice. My experiments confirm the doctrines (with some modifications) taught by SPALLANZINI, and ma ny of the most enlightened physiological writers. They are experiments made in the true spirit of enquiry, suggested by the very extraordinary case which gave me an opportunity of making them. I had no particular hypothesis to support; and I have therefore honestly recorded the result of each experiment exactly as it occurred.

The reader will perceive some slight seeming discrepancies, which he may find it difficult to reconcile; but he will recollect that the human machine is endowed with a vitality which modifies its movements in different states of the system, and probably produces some diversity of effects from the same causes.

I had opportunities for the examination of the interior of the stomach, and its secretions, which has never before been so fully offered to any one. This most important organ, its secretions and its operations, have been submitted to my observation in a very extraordinary manner, in a state of perfect health, and for years in succession. I have availed myself of the opportunity afforded by a concurrence of circumstances which probably can never again occur, with a zeal and perseverance proceeding from motives which my conscience approves; and I now submit the result of my experiments to an enlightened public, who I doubt not will duly appreciate the truths discovered, and the confirmation of opinions which before rested on conjecture.

I submit a body of facts which cannot be invalidated. My opinions may be doubted, denied, or approved, according as they conflict or agree with the opinions of each individual who may read them; but their worth will be best determined by the foundation on which they rest—the incontrovertible facts.

I avail myself of this opportunity to make my grateful acknowledgements to Doctor Joseph Lovell, Surgeon General of the United States' Army, (to whom I am under obligations for personal kindness and official exertions in affording facilities for prosecuting the experiments;)—to Professors Silliman, Knight, Ives and Hubbard, of Yale College, Dunglison, of the Virginia University, and Sewall, Jones, Henderson and Hall, of Columbian College, for their unsolicited friendship; for the interest which they have taken in the experiments, and for the generous encouragement which they have given to the proposed publication. To Doctor Samuel Beaumont, of Plattsburgh, N. Y. I am particularly indebted for the assistance which he has rendered me in arranging and preparing my notes for the press.

INTRODUCTION.

THE experiments which follow were commenced in 1825, and have been continued, with various interruptions, to the present time, (1833.) The opportunity for making them was afforded to me in the following way.

Whilst stationed at Michillimackinac, Michigan Territory, in 1822, in the military service of the United States, the following case of surgery came under my care and treatment.

Alexis St. Martin, who is the subject of these experiments, was a Canadian, of French descent, at the above mentioned time about eighteen years of age, of good constitution, robust and healthy. He had been engaged in the service of the American Fur Company, as a voyageur, and was accidentally wounded by the discharge of a musket, on the 6th of June, 1822.

The charge, consisting of powder and duck shot, was received in the left side of the youth, he being at a distance of not more than one yard from the muzzle of the gun. The contents entered posteriorly, and in an oblique direction, forward and inward, literally blowing off integuments and muscles of the size of a man's hand, fracturing and carrying away the anterior half of the sixth rib, fracturing the fifth, lacerating the lower portion of the left lobe of the lungs, the diaphragm, and perforating the stomach.

The whole mass of materials forced from the musket, together with fragments of clothing and pieces of fractured ribs, were driven into the muscles and cavity of the chest.

I saw him in twenty-five or thirty minutes after the accident occurred, and, on examination, found a portion of the lung, as large as a Turkey's egg, protruding through the external wound, lacerated and burnt; and immediately below this, another protrusion, which, on further examination, proved to be a portion of the stomach, lacerated through all its coats, and pouring out the food he had taken for his breakfast, through an orifice large enough to admit the fore finger.

In attempting to return the protruded portion of the lung, I was prevented by a sharp point of the fractured rib, over which it had caught by its membranes; but by raising it with my finger, and clipping off the point of the rib, I was able to return it into its proper cavity, though it could not be retained there, on account of the incessant efforts to cough.

The projecting portion of the stomach was nearly as large as that of the lung. It passed through the lacerated diaphragm and external wound, mingling the food with the bloody mucus blown from the lungs.

After cleansing the wound from the charge and other extraneous matter, and replacing the stomach and lungs as far as practicable, I applied the carbonated fermenting poultice, and kept the surrounding parts constantly wet with a lotion of muriate of ammonia and vinegar; and gave internally the aq. acet am. with camphor, in liberal quantities.

Under this treatment a strong reaction took place in about twenty-four hours, accompanied with high arterial excitement, fever, and marked symptoms of inflammation of the lining membranes of the chest and abdomen, great difficulty of breathing, and distressing cough.

He was bled to the amount of eighteen or twenty ounces, and took a cathartic. The bleeding reduced the arterial action, and gave relief. The cathartic had no effect, as it escaped from the stomach through the wound.

On the 5th day a partial sloughing of the integuments and muscles took place. Some of the protruded portions of the lung, and lacerated parts of the stomach, also sloughed, and left a perforation into the stomach, plainly to be seen, large enough to admit the whole length of my fore-finger into its cavity; and also a passage into the chest, half as large as my fist, exposing to view a part of the lung, and permitting the free escape of air and bloody mucus at every respiration.

A violent fever continued for ten days, running into a typhoid type, and the wound became very fœtid.

On the eleventh day, a more extensive sloughing took place, the febrile symptoms subsided, and the whole surface of the wound assumed a healthy and granulating appearance.

For seventeen days, all that entered his stomach by the œsophagus, soon passed out through the wound; and the only way of sustaining him was by means of nutricious injections per anus; until compresses and adhesive straps could be applied so as to retain his food. During this period no alvine evacuations could be obtained, although cathartic injections were given, and various other means were adopted to promote them.

In a few days after firm dressings were applied, and the contents of the stomach retained, the bowels became gradually excited, and, with the aid of cathartic injections, a very hard, black, fœtid stool was procured, followed by several similar ones; after which the bowels became quite regular, and continued so.

The cataplasms were continued until the sloughing was completed, and the granulating process fully established; and were afterwards occasionally resorted to, when the wound became ill conditioned. The aq. acet. am. with camphor was also continued for several weeks, in proportion to the febrile symptoms, and the fœtid condition of the wound.

No sickness, nor unusual irritation of the stomach, not even the slightest nausea, was manifest during the whole time; and after the fourth week, the appetite became good, digestion regular, the alvine evacuations natural, and all the functions of the system perfect and healthy.

By the adhesion of the sides of the protruded portions of the stomach to the pleura costalis and the external wound, a free exit was afforded to the contents of that organ, and effusion into the abdominal cavity was thereby prevented.

Cicatrization and contraction of the external wound commenced on the fifth week; the stomach became more firmly attached to the pleura and intercostals, by its external coats; but showed not the least disposition to close its orifice; this (the orifice) terminated as if by a natural boundary, and left the perforation, resembling, in all but a sphincter, the natural anus, with a slight prolapsus.

Whenever the wound was dressed the contents of the stomach would flow out, in proportion to the quantity recently taken. If the stomach happened to be empty, or nearly so, a partial inversion would take place, unless prevented by the application of the finger. Frequently in consequence of the derangement of the dressing, the inverted part would be found of the size of a hen's egg. No difficulty, however, was experienced in reducing it by gentle pressure with the finger, or a sponge wet with cold water, neither of which produced the least pain.

In the seventh week, exfoliation of the ribs, and a separation of their cartilaginous ends, began to take place.

The sixth rib was denuded of its periosteum for about two inches from the fractured part, so that I was obliged to amputate it about three or four inches from its articulation with the rib. This I accomplished by dissecting back the muscles, securing the intercostal artery, and sawing off the bone with a very fine narrow saw, made for the purpose, introduced between the ribs, without injury to the neighbouring parts. Healthy granulations soon appeared, and formed soundly over the amputated end. About half the inferior edge of the fifth rib exfoliated, and sepa rated from its cartilage.

After the removal of these pieces of bone, I attempted to contract the wound, and close the perforation of the stomach, by gradually drawing the edges together with adhesive straps, laid on in a radiated form.

The circumference of the external wound was at least twelve inches, and the orifice in the stomach nearly in the centre, two inches below the left nipple, on a line drawn from this to the point of the left ilium.

To retain his food and drinks I kept a compress and tent of lint, fitted to the shape and size of the perforation, and confined there by adhesive straps.

After trying all the means in my power for eight or ten months to close the orifice, by exciting adhesive inflammation in the lips of the wound, without the least appearance of success, I gave it up as impracticable in any other way than that of incising and bringing them together by sutures; an operation to which the patient would not submit.

By the sloughing of the injured portion of the lung, a cavity was left as large as a common sized teacup, from which continued a copious discharge of pus for three months, when it became filled with healthy granulations, firmly adhering to the pleura, and soundly cicatrized over that part of the wound.

Four months after the injury was received, an abscess formed about two inches below the wound, nearly over the cartilaginous ends of the first and second false ribs, very painful, and extremely sore, producing violent symptomatic fever. On the application of an emollient poultice it pointed externally. It was then laid open to the extent of three inches, and several shot and pieces of wad extracted. After which a gum-elastic bougie could be introduced three or four inches in the longitudinal direction of the ribs, towards the spine. Great pain and soreness extended from the opening of the abscess, along the track of the cartilaginous ends of the fale ribs, to the spine, with a copious discharge from the sinus.

In five or six days there came away a cartilage, one inch in length. In six or seven days more, another, an inch and a half long; and in about the same length of time, a third, two inches long, were discharged. And they continued to come away every five or six days, until *five* were discharged from the same opening, the last three inches in length. They were all entire, and evidently separated from the false ribs.

The discharge, pain and irritation, during the four or five weeks these cartilages were working out, greatly reduced the strength of the patient, produced a general febrile habit, and stopped the healing process of the original wound.

Directly after the discharge of the last cartilage, inflammation commenced over the lower end of the sternum, which by the usual applications, terminated in a few days in a large abscess, and from which, by laying it open two inches, I extracted another cartilage, three inches in length. The inflammation then abated; and in a day or two another piece came away, and the discharge subsided.

To support the patient under all these debilitating circumstances, I administered wine, with diluted muriatic acid, and thirty or forty drops of the tincture of assafœtida, three times a day; which appeared to produce the desired effect, and very much improved the condition of the wound.

On the third of January, 1823, I extracted another cartilage from the opening over the sternum, an inch and a half long; and on the fourth another, two inches and a half in length, an inch broad at one end, and narrowing to less than half an inch at the other. This must have been the ensiform cartilage of the sternum. After this the sinus closed, and there was no return of inflammation.

From the month of April, 1823, at which time he had so far recovered as to be able to walk about and do light work, enjoying his usual good appetite and digestion, he continued with me, rapidly regaining his health and strength.

By the 6th of June, 1823, one year from the time of the accident, the injured parts were all sound, and firmly cicatrized, with the exception of the aperture in the stomach and side. This continued much in the same situation as it was six weeks after the wound was received. The perforation was about two and a half inches in circumference, and the food and drinks constantly exuded, unless prevented by a tent, compress and bandage.

From this time he continued gradually to improve in health and strength, and the newly formed integuments over the wound became firmer and firmer. At the point where the lacerated edges of the muscular coat of the stomach and intercostal muscles met and united with the cutis vera, the *cuticle* of the external surface and the *mucous membrane* of the stomach *ap-*

proached each other very nearly. They did not unite, like those of the lips, nose, &c., but left an intermediate marginal space, of appreciable breadth, completely surrounding the aperture. This space is about a line wide; and the cutis and nervous papillæ are unprotected, as sensible and irritable as a blistered surface abraded of the cuticle. This condition of the aperture still continues, and constitutes the principal and almost only cause of pain or distress experienced from the continuance of the aperture, the introduction of instruments, &c. in the experiments, or the exudation of fluids from the gastric cavity.

Frequent dressings with soft compresses and bandages were necessarily applied, to relieve his suffering and retain his food and drinks, until the winter of 1823-4. At this time, a small fold or doubling of the coats of the stomach appeared, forming at the superior margin of the orifice, slightly protruding, and increasing till it filled the aperture, so as to supersede the necessity for the compress and bandage for retaining the contents of the stomach. This valvular formation adapted itself to the accidental orifice, so as completely to prevent the efflux of the gastric contents when the stomach was full, but was easily depressed with the finger.

In the spring of 1824 he had perfectly recovered his natural health and strength; the aperture remained; and the surrounding wound was firmly cicatrized to its edges.

In the month of May, 1825, I commenced my first series of gastric experiments with him, at Fort Mackinac, Michigan Territory. In the month of

June following, I was ordered to Fort Niagara, N. Y. where, taking the man with me, I continued my experiments until August. Part of these experiments were published in 1826, in the 29th number of the Philadelphia " Medical Recorder," conducted by Doctor Samuel Calhoun. About this time, (August 1825) I took St Martin with me to Burlington, Vermont, and from thence to Plattsburgh, New-York. From the latter place, he returned to Canada, his native place, without obtaining my consent.

Being unable to ascertain the place of his resort, I gave him up as a lost subject for physiological experiments, and returned to my post at the west again. I did not, however, remit my efforts to obtain information of his place of residence and condition.

He remained in Canada four years, during which period he married, and became the father of two children; worked hard to support his family; and enjoyed robust health and strength. In 1825, as he has informed me, he engaged with the Hudson Bay Fur Company, as a voyageur to the Indian country. He went out in 1827, and returned in 1828; and subsequently laboured hard to support his family until 1829.

Accidentally learning about this time where he was, and that he enjoyed perfect health, I made arrangements with the agents of the American Fur Company, who annually visit Canada for the purpose of procuring voyageurs, to find and engage him for my service, if practicable. After considerable difficulty, and at great expense to me, they succeeded in engaging him, and transported him from Lower Canada, with his wife and two children, to me, at Fort Crawford, Prairie du Chien, Upper Mississippi, a distance of nearly two thousand miles, in August, 1829. His stomach and side were in a similar condition as when he left me in 1825. The aperture was open, and his health good.

He now entered my service, and I commenced another series of experiments on the stomach and gastric fluids, and continued them, interruptedly, until March, 1831. During this time, in the intervals of experimenting, he performed all the duties of a common servant, chopping wood, carrying burthens, &c. with little or no suffering or inconvenience from his wound. He laboured constantly, became the father of more children, and enjoyed as good health and as much vigour as men in general. He subsisted on crude food, in abundant quantities, except when on prescribed diet, for particular experimental purposes, and under special observance.

In the spring of 1831 circumstances made it expedient for him to return with his family from Prairie du Chien to Lower Canada again. I relinquished his engagements to me for the time, on a promise that he would return when required, and gave him an outfit for himself, wife and children. They started in an open canoe, via the Mississippi, passing by St Louis, Mo.; ascended the Ohio river; then crossed the state of Ohio, to the Lakes; and descended the Erie, Ontario, and the River St. Lawrence, to Montreal, where they arrived in June. He remained in Canada with his family until October, 1832, in good health, and at hard labour. He was in the midst of the cholera epidemic, at the time it prevailed, and passed through Canada, and withstood its ravages with impunity, while hundreds around him fell sacrifices to its fatal influence.

In November, 1832, he again engaged himself to me for twelve months, for the express purpose of submitting to another series of experiments. He joined me at Plattsburgh, N. Y., and travelled with me to the city of Washington, where, with the facilities afforded by the head of the Medical Department, the experiments were continued upon him from November, 1832, to March, 1833.

During the whole of these periods, from the spring of 1824 to the present time, he has enjoyed *general* good health, and perhaps suffered much less predisposition to disease than is common to men of his age and circumstances in life. He has been active, athletic and vigorous; exercising, eating and drinking like other healthy and active people. For the last four months, he has been unusually plethoric and robust, though constantly subjected to a continued series of experiments on the interior of the stomach; allowing to be introduced or taken out at the aperture different kinds of food, drinks, elastic catheters, thermometer tubes, gastric juice, chyme, &c., almost daily, and sometimes hourly.

Such have been this man's condition and circumstances for several years past; and he now enjoys the most perfect health and constitutional soundness, with every function of the system in full force and vigour.

Mode of extracting the Gastric Juice.—The usual method of extracting the gastric juice, for experiment, is by placing the subject on his right side, depressing the valve within the aperture, introducing a gum-elastic tube, of the size of a large quill, five or six inches into the stomach, and then turning him on the left side, until the orifice becomes dependent. In health, and when free from food, the stomach is *usually* entirely empty, and contracted upon itself. On introducing the tube, the fluid soon begins to flow, first by drops, then in an interrupted, and sometimes in a short continuous stream. Moving the tube about, up and down, or backwards and forwards, increases the discharge. The quantity of fluid ordinarily obtained is from four drachms to one and a half or two ounces, varying with the circumstances and condition of the stomach. Its extraction is generally attended by that peculiar sensation at the pit of the stomach, termed sinking, with some degree of faintness, which renders it necessary to stop the operation. The usual time of extracting the juice is early in the morning, before he has eaten, when the stomach is empty and clean.

On laying him horizontally on his back, pressing the hand upon the hepatic region, agitating a little, and at the same time turning him to the left side, bright yellow bile appears to flow freely through the pylorus, and passes out through the tube. Sometimes it is found mixed with the gastric juice, without this operation. This is, however, seldom the case, unless it has been excited by some other cause.

The chymous fluids are easily taken out by depressing the valve within the aperture, laying the hand over the lower part of the stomach, shaking a little, and pressing upwards. In this manner, any quantity necessary for examination and experiment can be obtained.

Valve—The valve mentioned above, is formed by a slightly inverted portion of the inner coats of the stomach, fitted exactly to fill the aperture. Its principal and most external attachment is at the upper and posterior edge of the opening. Its free portion hangs pendulous, and fills the aperture when the stomach is full, and plays up and down, simultaneously with the respiratory muscles, when empty.

On pressing down the valve when the stomach is full, the contents flow out copiously. When the stomach is nearly empty, and quiescent, the interior of the cavity may be examined to the depth of five or six inches, if kept distended by artificial means ; and the food and drinks may be seen entering it, if swallowed at this time, through the ring of the æsophagus. The perforation through the walls of the stomach, is about three inches to the left of the cardia, near the left superior termination of the great curvature. When entirely empty, the stomach contracts upon itself, and sometimes forces the valve through the orifice, together with an additional portion of the mucous membrane, which becomes completely inverted, and forms a tumour as large as a hen's egg. After lying on the left side, and sleeping a few hours, a still larger portion protrudes, and spreads out over the external integuments, five or six inches in circumference, fairly exhibiting the natural rugæ, villous membrane, and mucous coat, lining the gastric cavity. This appearance is almost invariably exhibited in the morning, before rising from his bed.

PLATTSBURGH, 1833.

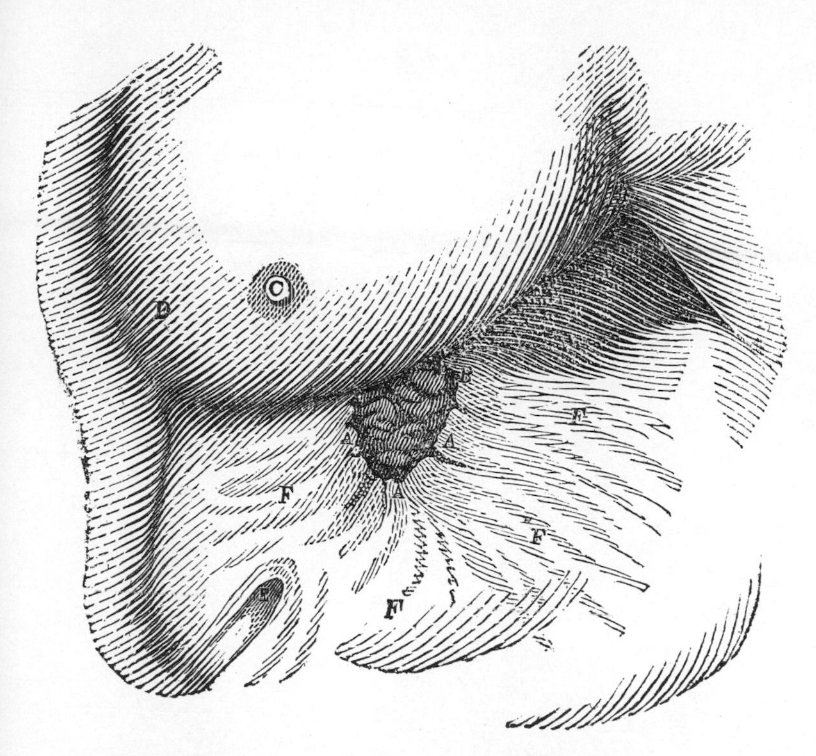

This engraving represents the ordinary appearance of the left breast and side, the aperture filled with the valve; the subject in an erect position.

A A A The circumference and edge of the aperture, within which is seen the valve.

B The attachment of the valvular portion of the stomach to the superior part of the aperture.

C The nipple.

D The anterior portion of the breast.

E The scar where the opening was made with the scalpel, and the cartilages taken out.

F F F F Cicatrice of the original wound, around the aperture.

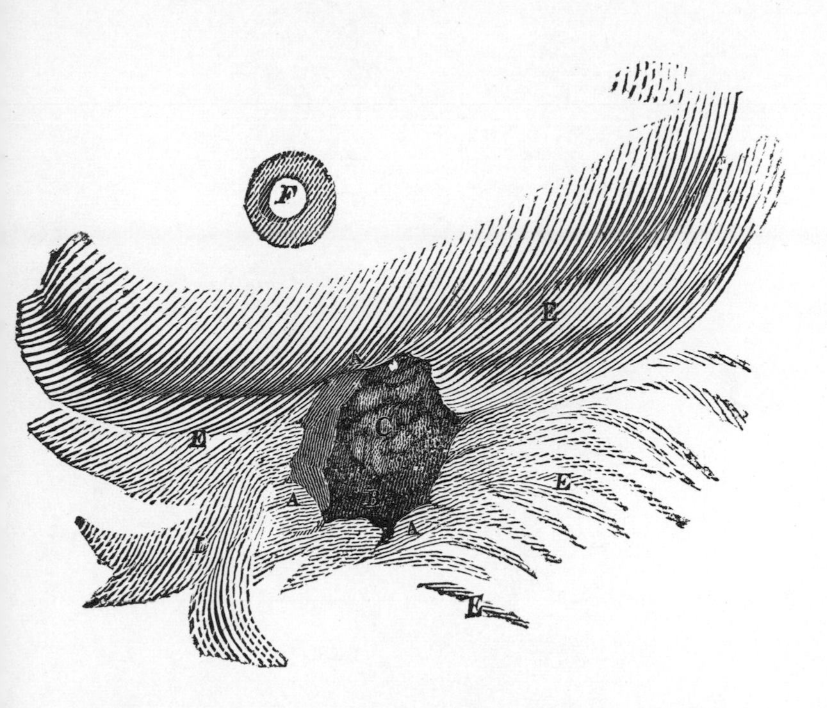

This engraving represents the appearance of the aperture with the valve depressed.

A A A Edges of the aperture through the integuments and intercostals, on the inside and around which is the union of the lacerated edges of the perforated coats of the stomach with the intercostals and skin.

B The cavity of the stomach, when the valve is depressed.

C Valve, depressed within the cavity of the stomach.

E E E E Cicatrice of the original wound.

F The nipple.

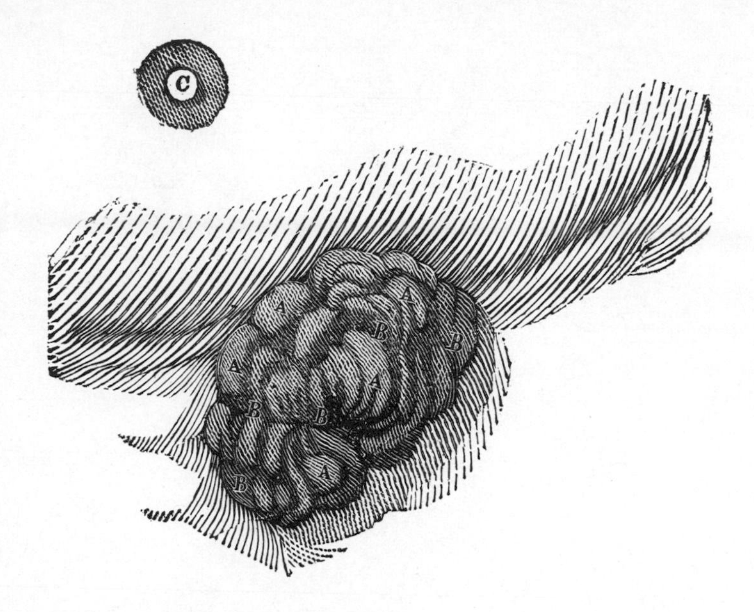

This engraving represents a portion of the stomach prolapsed through the aperture, with the inner surface inverted, and spread out over the integuments of the side.

A A A A A Folds or rugæ of the inner coats of the stomach.

B B B B Interstices, filled with mucous substance.

C The nipple.

PRELIMINARY OBSERVATIONS.

I do not design, in the following remarks, to present any thing like a systematic treatise on digestion. Works of this kind, treating of the subject both physiologically and pathologically, have so multiplied of late, as to render an attempt on my part, entirely a work of supererogation, even if I believed myself qualified for the task. I consider myself but a humble inquirer after truth—a simple experimenter. And if I have been led to conclusions opposite to the opinions of many who have been considered the great luminaries of physiology, and, in some instances, from all the professors of this science, I hope the claim of sincerity will be conceded to me, when I say that such difference of opinion has been forced upon me by the convictions of experiment, and the fair deductions of reasoning.

I shall not attempt an anatomical description of the organs of digestion, for the reasons given above. In a work professedly elementary, such descriptions are essential. The medical profession are acquainted with these organisms. The general reader, if he have a wish for information of this kind, is referred to anatomical authors generally, or to the physiological writings of Richerand, Broussais, Magendie, Bostock, Fordyce, Paris, Jackson and Dunglison, the last of which, as containing the sum of what has been taught in the schools on the subject of physiology generally, arranged in a clear and distinct manner, and with the assistance of numerous plates, is well worthy his perusal.

With a view to comment on my experiments, and to elucidate my opinions on the subject of digestion, I shall divide my observations into the following heads:

Section 1st. *Of Aliment.* Section 2d. *Of Hunger and Thirst.* Section 3d. *Of Satisfaction and Satiety.* Section 4th. *Of Mastication, Insalivation and Deglutition.* Section 5th. *Of Digestion by the Gastric Juice.* Section 6th. *Of the appearance of the Villous Coat, and of the Motions of the Stomach.* Section 7th. *Of Chylification. and Uses of the Bile and Pancreatic Juice*

SECTION I.

Of Aliment.

Man is said to be an *omnivorous* animal, destined to procure his food from both the animal and vegetable kingdoms. The inhabitant of temperate climates is unquestionably so. It would be interesting to ascertain by experiment whether he could be sustained by habit, from infancy, on the productions of either of these grand divisions. If the result should be favourable to the demonstration of this proposition, though it might still more unsettle the opinions of physiologists, it would be an evidence of this truth, that man is a creature of habit and circumstance, carrying about him the effects of primeval disobedience, destined not only to earn his food by his own exertions, but to partake of such as the climate in which he resides may supply to him. Approximating to this are the habits of people of different quarters of the world—those of Asia, who live almost exclusively on vegetable and farinaceous food, and those of the northern regions of America, who derive their food principally from fish, oil and flesh.

Other substances have sometimes been used as aliment; and Professor Dunglison mentions, on the authority of Humboldt, that the Ottomaques, a tribe of Indians of South America, are in the habit of using "an unctious earth, or a species of pipe clay," as an article of diet. Whether nutriment can be supplied by such articles alone, is extremely problematical. In all countries, some persons are found who are in the constant habit of eating large quantities of clay, chalk, slate stone, &c. Such practices may be regarded as evidence, if not of a diseased, at least of a vitiated appetite; though it often happens that alkaline and absorbent substances are used medicinally with advantage, particularly where much acidity of the stomach prevails.

As it respects the inhabitants of Europe and their American descendants, as well as most other natives of temperate climates, it is well known that they derive their nourishment from both the animal and vegetable kingdoms.

The facility of digestion of different articles of diet, and the quantity of nutrient principles which they contain, have been subjects of some discrepance of opinion among physiologists. They have, however, settled down into a belief, probably as near the truth as practicable, that animal food is more readily assimilated, and affords more nutrition in a given quantity, than vegetable or farinaceous food.

Animal food has been divided into fibrine, gelatine and albumen, and a comparison drawn between their degrees of digestibility. But it will occur to every one at all acquainted with the subject, that almost every portion of animal food contains an admixture of all these principles, and it is consequently very difficult to come to a correct conclusion. The truth is, there can be no general rule on this subject. The facility of digestion is modified by so many circumstances, as health, disease, idiosyncracy, habit, and preparation of food, that a rule which would apply in one case would be incorrect in another. It depends more upon other distinctions than upon those relating to the chemical composition of the food. Albumen, (one of these chemical divisions,) if taken into the stomach, either very slightly or not at all coagulated, is perhaps as rapidly chymified as any article of diet we possess. If perfectly formed into hard coagulæ, by heat or otherwise, and swallowed in large solid pieces, it experiences a very protracted digestion. The reason is obvious. In the first case the albumen becomes finely coagulated, and divided in the stomach; in the second, it is less susceptible of subdivision from its hardness. Fibrine and gelatine are affected in the same way. If tender and finely divided, they are disposed of readily; if in large and solid masses, digestion is proportionably retarded. Minuteness of division and tenderness of fibre are the two grand essentials for speedy and easy digestion. By referring to my experiments, it will be seen that those articles of diet which were submitted to the action of the gastric juice, either artificially, when out of the stomach, or in the stomach, by natural process, were dissolved in proportion to the fineness of their division or their solidity—the one rapidly, and the other slowly.

The digestion of animal and vegetable diet requires the same process, though one may afford a larger proportion of the nutrient principle than the other. Generally speaking, vegetable aliment requires more time, and probably greater powers of the gastric organs, than animal. Its digestibility is, however, dependant upon the same laws as those that govern the solution of animal food; and it is facilitated by division and tenderness.

The ultimate principles of nutriment are probably always the same, whether obtained from animal or vegetable diet. It was said by Hippocrates, that "there are many kinds of aliments, but that there is at the same time but one aliment." This opinion has been contested by most modern physiologists; but I see no reason for scepticism on this subject. Some imperfect experiments which I instituted on the operations of the hepatic and pancreatic juices, and which will be found in a subsequent part of this volume, tend to throw some light on the subject. Chyme was submitted to the action of these fluids, and they invariably produced similar effects. A fluid was separated, varying slightly in colour, but of the same apparent consistence and identity; and was increased or lessened in proportion to the quality of the food of which the chyme was formed. Whether this fluid was or was not imperfectly formed chyle, is a matter of opinion only. The circulating fluids of the system are always nearly the same, in health, and that which goes to supply and replenish them, should consequently possess the same invariable properties. Chyle, after its separation in the intestines, is probably further changed and perfected by the action of the lacteal absorbents and sanguiferous vessels, before it is completely assimilated. Chyme, from which this nutrient principle is obtained, is a compound of gastric juice and aliment. It may be regarded as a *gastrite* of whatever it is combined with, varied according to the kind of aliment used. The perfect chyle, or assimilated nutriment, probably contains the elements of all the secretions of the system; such as bone, muscle, mucus, saliva, gastric juice, &c. &c., which are separated by the action of the glands, the sanguiferous and other vessels of the system.

The action of the stomach, and its fluids, on aliment, is believed to be *sui generis*, invariably the same, in health, on all kinds. And yet it is contended by Paris, and obliquely hinted by some other modern physiologists, that as animal food "posseses a composition analogous to that of the structure it is designed to supply," it "requires little more than division and depuration," &c. It is singular that sensible men, and men of science, can allow themselves to be led to such erroneous conclusions, and will not perceive a simplicity and *uniformity* in the process of digestion, as well as in all the other operations of nature. That the active solvent of the stomach should produce the same effect on all alimentary substances, is no more wonderful than that caloric should liquefy all kinds of matter. In both cases it only requires a longer or shorter continuance, or more or less concentrated action, of the agent, to produce the same effect. If animal food is only to be divided and depurated, blood, which is an elementary part of the body, would require no change in the stomach. But it is perfectly idle to talk in this way. The most innutritious vegetable and the most animalized substance, require the same action of the gastric solvent, as the reader will find amply demonstrated in the following experiments. It is true that one may be disposed of with ease, and the other with difficulty; but this is not always, nor indeed often, in a direct ratio to their respective proportions of nutrient principles. An innutritious diet may be disposed of as easily, the circumstances of divisibility and tenderness of fibre being equal, as a nutritious one. I do not believe that the one requires a more "complicated series of decompositions and recompositions" than the other; nor that the chyle from animal aliment is more highly "animalized" than that from the poorest diet we possess. The "digestive fever," or the excitement that follows the digestion of animal food, is the effect, not of a different *kind* of stimulus, but of the introduction of a greater *quantity* of chyle, or the nutritive principle of food, into the circulating fluids. It excites the system precisely in the same manner as ardent spirits, or other stimulus does, with the exception, that its effects are more permanent.

The quantity of nutriment required by different individuals, is as various as the individuals who partake of it. As a general rule, it may be said that persons who do not exercise much, require less nutritious diet than those who are in the habit of constant labour. What would be a natural supply in one, would be excess in another. With labouring persons, much of the excess is carried off by perspiration; and probably a great deal of nervous energy is wasted by laborious occupations, which requires to be replenished by the nutrient principles of aliment. This is a subject, however, on which we can only offer conjecture; for it is difficult to argue on a point of which we know so little. Young people who are growing, require more nutriment in proportion to their size, than those who have arrived at adult age.

The quality of nutriment is a matter of considerable importance in dietetic regulations. Bulk is, perhaps, nearly as necessary to the articles of diet as the nutrient principle. They should be so managed that one should be in proportion to the other. Too highly nutritive diet is probably as fatal to the prolongation of life and health, as that which contains an insufficient quantity of nutriment. It has been ascertained that carnivorous animals will not live on highly concentrated food alone. Dogs fed on oil or sugar, which are both converted by the digestive organs almost entirely into chyle, are found to become diseased, and die in a few weeks. The inference drawn by Paris,* that it merely "proves that an animal cannot be supported on highly concentrated aliment alone," no doubt, is a correct one; though opposed to the opinion of Magendie, the author of the experiments, who infers that death proceeds from the want of azote in these articles of diet, and that life cannot be supported on non-azotïzed aliment.

* Paris On Diet, p 72.

The following articles of the *materia alimentaria* have, in the course of these experiments, been submitted to the action of the stomach and the gastric fluids. I have attempted, in this table, to approximate towards a comparison of the digestibility of the several articles there mentioned. Precision, as to minutes, has not been attended to. When digestion has been accomplished two or three minutes either before or after a certain number of hours and quarters, I have set down the quarter to which it approached the nearest.

In a subsequent part of this volume a more particular and minute detail will be found, both of natural and artificial digestion.

Explanation of the abbreviations in the following Table.—br'd, for bread—veg., for vegetables—mod., for moderate—inc., for increased—susp'd, for suspended —dur'g, for during—exp't, for experiment—h'rd bld, for hard boiled—h., for hour—m., for minute, &c.— The figures denote the time of digestion, under the circumstances mentioned at the head of the column.

Articles of Diet.	mode of cooking	meal	With br'd or veg. or both — exercise. mod.	exercise. inc.	rest	Remarks.
			h. m.	h. m.	h. m.	
Tripe, soused	fried	breakfast	1 00			
Pig's feet, do.	boiled	..	1 00			
Venison steak, fresh,	broil'd	..	1 35			
Codfish, dry,	boiled	dinner	2 00			
Bread and Milk,	cold	..	2 00			
Turkey,	roast'd	..	2 30			
Goose, wild,	roast'd	..	2 30			
Pig, young,	..	..	2 30			
Hash, meat & veg.	warm	breakfast	2 30			
Oysters,	raw	dinner	2 45	-	-	oyster susp'd in stom. dur'g exp't.
do.	stewed	..	3 30	-	-	Nothing but a little dry bread or cracker taken at these meals.
do.	raw	breakfast	3 00	-	-	
do.	..	dinner	3 00	-	-	
do.	stewed	..	3 30	-	-	
Beef, fresh, fat & lean	roast'd	dinner	3 30			
do.	..	..	3 00			
do.	..	breakfast	2 45			
do.	broil'd	..	3 00			

Articles of Diet.	mode of cooking	meal	With br'd or veg. or both: exercise, mod.	exercise, inc.	rest	Remarks.
			h. m.	h. m.	h. m.	
Beef, fresh, fat & lean	broil'd	breakfast			3 45	
do.	. .	. .	-	3 30	-	Exercised till fatigued.
do.	boiled	. .	4 00		-	Morbid appearance of stomach.
do.	. .	dinner	-	3 30		
do.	boiled	breakfast	3 38	-	-	Large proportion of fat.
do.	. .	supper	-	-	4 00	do.
do.	. .	breakfast	-	-	4 30	do. & in recumbent position.
do.	. .	dinner	3 30			
do.	. .	. .	-	-	4 00	
do.	. .	breakfast	-	-	4 15	
do.	. .	. .	3 30			
do.	. .	. .	-	-	4 15	
do. salted,	. .	dinner	5 30			
do. do.	. .	. .	3 30			
Pork, recently salt'd	. .	breakfast	5 15			
do. .	. .	. .	4 30			
do. .	. .	. .	5 15			Became angry during experiment
do. .	. .	. .	6 00			Unusually full meal.

Articles of Diet.	mode of cooking	meal	With br'd or veg or both: exercise mod.	exercise inc.	rest	Remarks.
			h. m.	h. m.	h. m.	
PORK, recently salt'd	boiled	breakfast	4 30			
do. .	. .	. .	4 30			
do. .	. .	. .	4 30			
do. .	. .	dinner	4 30			
do. .	. .	breakfast		4 00		
do. .	. .	dinner		3 30		
do. fresh,	roast'd	dinner	6 30	-	-	Unusually full meal.
do. steak,	broil'd	. .	3 15			
do. do.	. .	breakfast	4 30			
MUTTON, fat & lean	roast'd	dinner	3 15			
do. .	broil'd	breakfast		3 00		
do. .	. .	. .	3 30			
do. .	. .	. .	4 30			Morbid appearance of stomach.
do. .	. .	dinner	4 00			
do. .	. .	breakfast	4 30	-	-	Full meal, coarsely masticated.
EGGS, .	hr'd bld	. .	3 30	-	-	Bread or bread & coffee, no vegetables used with the eggs.
do. .	soft bld	. .	3 00			
do. .	hr'd bld	dinner	5 30	-	-	Morbid appearance of stomach.

Articles of Diet.	mode of cooking	meal	With br'd or veg. or both: exercise mod.	exercise inc.	rest	Remarks.
			h. m.	h. m.		
EGGS, .	h'rd bld	breakfast	3 30	.	.	Morbid appearance of stomach.
do. .	soft bld	dinner	3 00			
SAUSAGE, .	broil'd	breakfast	3 30	-	-	With soft boiled eggs.
do. -	. .	dinner	3 00			
do. -	fried	breakfast	4 00	-	-	Muslin bag containing same kind of diet, susp'd dur'g these exp'ts—morbid appearance of stomach also.
do. -	. .	. .	5 00	-	-	
do. -	broil'd	. .	3 30			
do. -	. .	. .		4 15		Full meal—severe exercise
FOWLS, (hens)	boiled	. .	4 00	-	-	With bread and coffee.
do. -	. .	dinner	4 00	-	-	With bread and water.
do. -	. .	. .	4 00	-	-	do. do.
VEAL, fresh, -	broiled	breakfast	4 00	-	-	Muslin bag susp'd in stomach.
do. -	. .	dinner	4 00			
do. -	. .	breakfast	4 00			
do. -	. .	dinner	4 45	-	-	Morbid appearance of stomach.
do. -	. .	breakfast		3 45		
do. -	. .	dinner	4 30			
do. -	. .	breakfast	5 30	-	-	Morbid appearance of stomach.

	h. m.
Soup, made of fresh muscular fibre of beef, and vegetables,	4 00
do. made of the hock, with vegetables,	4 15
Bread, buttered, for breakfast, with coffee [morbid appearance of stomach]	4 15
do. buttered, for breakfast, with coffee	3 45
do. dry, for breakfast, with coffee,	4 00
do. dry, for dinner, with dry mashed potatoes	3 45

This Table is far from being complete. The experiments from which it has been formed, were made principally with the view of demonstrating other important principles connected with the subject of digestion. The only way of ensuring minuteness and accuracy as to the comparative digestibility of different kinds of diet, would be to try the effect of the gastric juice, in a series of experiments, first on one article of diet, and then on another, repeating and adapting them to meet all the various conditions of the stomach, and the vicissitudes and irregularities of the system, until the whole range should be completed—a Herculean task, which it would take years to accomplish. In the above table, the time is counted from the reception of the meal of various articles to the chymification of the whole: hence the conclusions are frequently indefinite, some of the articles being sooner disposed of than others. For instance, if a dinner be eaten of venison steak and fat pork, the time of digestion of the whole quantity would, in all probability, be twice as long as if venison had been used alone. Oily substances are digested with great difficulty, and the fat of all meats is converted into oil in the stomach before it is digested. Chymification is most readily effected on solid food, or rather on a soft solid, which is easily divisible into shreds or small particles. Such is particularly the character of venison, which is ascertained to be one of the most digestible of substances. The qualities of looseness of texture and susceptibility of division belong to most of those wild meats and game which are generally acknowledged to be easy of digestion. Beef and mutton, of a certain age, possess similar qualities.

The opinion advanced by Paris, *that the flesh of wild animals is more dense than the domesticated, does not correspond with the experience of those who are well acquainted with the former. Although, on making a section of wild flesh, such appearance may be indicated, yet the fibres are found to be more easily separated by mastication, or other force, and are generally tender ; at least, such is the case with the flesh of those animals that are considered luxuries by the epicure. Compare, for example, the flesh of the wether and the deer, animals which have a near correspondence in their habits, and the difference will be very obvious.

The digestibility of most meats is improved by incipient putrefaction, sufficient to render the muscular fibre slightly tender.

Vegetables are generally slower of digestion than meats and farinaceous substances, though they sometimes pass out of the stomach before them, in an undigested state. Crude vegetables, by some law of the animal economy, not well understood, are allowed, even when the stomach is in a healthy state,

Paris On Diet, p. 72.

sometimes to pass the pyloric orifice, while other food is retained there to receive the solvent action of the gastric juice. This may depend upon their comparative indigestibility; for it is well known that cathartic medicines, various fruits, seeds, &c. which operate as laxatives, are not digested; are incapable of being retained in the stomach; and pass rapidly through the intestinal tube. When such articles are in excess, they produce considerable derangement, and sometimes fatal consequences.

Vegetable, like animal substances, are more capable of digestion in proportion to the minuteness of their division, as I have before remarked, provided they are of a soft solid; and I cannot, therefore, concur in the opinion expressed by Paris,* that potatoes are better when only boiled so as to be rendered tender, and have their shape preserved, than when boiled to a "dry, insipid powder." They may be more palatable, and contain more nutriment; but they are not so easily affected by the gastric solvent. The difference is quite obvious on submitting parcels of this vegetable, in different states of preparation, to the operation of the gastric juice, either in the stomach or out of it. Boiled, or otherwise cooked to dryness, so as to be easily mashed, potatoes very readily become reduced to a chymous state, when submitted to the action of the gastric juice. When differently prepared, and only boiled so as to be rendered barely soft, moist and tenacious, with the shape preserved, entire pieces remain long undissolved in the stomach, and very slowly yield to the action of the gastric juice in vials

Paris On Diet, p. 75.

on the bath. Pieces of raw potato, when submitted to the operation of this fluid, in the same manner, almost entirely resist its action. Many hours elapse before the slightest appearance of digestion is observable, and this only upon the surface, where the external laminæ become a little softened, mucilaginous, and slightly farinaceous. Every physician, who has had much practice in the diseases of children, knows that partially boiled potatoes, when not sufficiently masticated, (which is always the case with children,) are frequently a source of colics and bowel complaints, and that large pieces of this vegetable pass the bowels untouched by digestion.

These remarks will apply, also, to most other vegetable aliment.

The variety of fish, which are generally used by the citizens of this country, may be regarded as easily susceptible of digestion. The lobster, crab, and some others of the testaceous tribe, are, perhaps, exceptions.

Solid food is sooner disposed of by the stomach than fluid, and its nutritive principles are sooner carried into the circulation. It has been observed, however, that the exhaustion from abstinence is quicker removed by liquid than solid aliment. This is undoubtedly true; and it may be accounted for on the ground of a general sympathy existing between the stomach and all the other parts of the body. It is only necessary, in proof of this fact, to appeal to the experience of almost every physician. The violent spasms, contortions, &c. affecting different and remote parts of the system, that sometimes supervene on the introduction of crude or indigestible food into the stomach, are pretty clear indications of the powerful sympathy that exists between it and other organs or apparatuses.

Condiments, particularly those of the spicy kind, are non-essential to the process of digestion, in a healthy state of the system. They afford no nutrition. Though they may assist the action of a debilitated stomach for a time, their continual use never fails to produce an indirect debility of that organ. They affect it as alcohol or other stimulants do—the *present* relief afforded is at the expense of *future* suffering. Salt and vinegar are exceptions, and are not obnoxious to this charge, when used in moderation. They both assist in digestion—vinegar, by rendering muscular fibre more tender—and both together by producing a fluid having some analogy to the gastric juice.

Drinks are nearly as essential to the animal system as the more substantial food. Though not subject to digestion, they enter into the circulation, and become important agents in the ultimate changes that are undergoing in the tissues of the organism. Simple water is, perhaps, the only fluid that is called for by the wants of the economy. The artificial drinks are probably all more or less injurious; some more so than others; but none can claim exemption from the general charge. Even coffee and tea, the common beverages of all classes of people, have a tendency to debilitate the digestive organs. Let any one who is in the habit of drinking either of these articles in a weak decoction, take two or three cups made very strong, and he will soon be aware of their injurious tendency. And this is only an *addition* to the *strength* of the narcotic he is in the constant habit of using. The whole class of alcoholic liquors, whether simply fermented, or distilled, may be considered as *narcotics*, producing very little difference in their ultimate effects on the system.

The injury which a constant use of wine is known to produce on some stomachs, has been sometimes attributed to the small quantity of tartaric acid which it contains. But it is not the cream of tartar that renders wine so deleterious to many stomachs. It is the acidity produced by the acetous fermentation of the saccharine matter contained in the wine, aided, perhaps, by the alcohol which is in a state of combination with it. Beer has the same effect on the same idiosyncracies, or diseased states of the stomach. Besides, both of these fluids are in a partial stage of acetous fermentation, which is consummated by the increase of temperature in the stomach.

It would be a task of great difficulty to designate the exact kind of diet that would, if generally adopted, be the most conducive to health and longevity. A considerable variety seems to be necessary to man, in a state of civilization. This want of variety is induced by long habit, which it would probably be unsafe to break through. Whether man was originally carnivorous or granivorous, is a question which we cannot solve, and perhaps it is not worth the attempt; at present he is both, and with his present mode of existence we have to do.

The *quantity* of aliment is probably of more importance than the *quality*, to ensure health. The system requires much less than is generally supplied to it. The stomach disposes of a definite quantity. If more be taken than the actual wants of the economy require, the residue remains in the stomach, and becomes a source of irritation, and produces a consequent abberration of function, or passes into the lower bowels in an undigested state, and extends to them its deleterious influence. Dyspepsia is oftener the effect of over eating and over drinking than of any other cause.

SECTION II.

Of Hunger and Thirst.

HUNGER is a painful sensation, referred to the region of the stomach. It is a kind provision of nature, designed to remind man, and other animated beings, of the necessity of replenishing the wastes of the system, as well as contributing to its growth. Much enquiry has been made on this subject, and many theories have been given to account for the phenomenon. It has been supposed by some, that the friction of the internal coats of the empty stomach was the cause of the sensation. This opinion is liable to several objections:—1st. A healthy stomach digests its contents in from one to three or four hours, and hunger is not usually experienced until some time after the latter period. If hunger be the effect of the friction of the parieties of the stomach, it ought to be experienced the moment that that organ has disposed of its contents. 2d. In nausea and vomiting, the stomach is brought into a situation, according to this theory, to experience the sensation of hunger; and yet we know how opposed it is to receiving any thing like food. 3d. In gastritis and fevers the sensation hardly ever occurs, though very little food shall have occupied the stomach for a long time—perhaps not for weeks. This organ, under such circumstances, is generally empty and irritable, yet the peculiar sensation in question hardly ever supervenes. Besides, hunger sometimes occurs when the stomach is partially or wholly filled. The potation of spirits or brandy and water, and some other indigestible substances of a liquid character, does not remove the sensation, although by this means the parieties of the stomach are as completely separated as by food.

It has also been suggested that the sensation of hunger is produced by the irritation of a quantity of gastric juice in the stomach, which, by its stimulus, excites the feeling. The principal objection to this doctrine is based upon the fact that the stomach contains no gastric juice, or, at any rate, but a very small quantity, in its empty state, or when aliment or other irritant is not present. Besides if it were true that it contained a quantity of the fluid, such fluid does not possess the power of producing any thing like irritation or inflammation of its coats. It is as innoxious to the stomach, as the blandest substance in nature. It exerts its influence on free aliment, but not on the living fibre.

By referring the sensation to "an energetic state of the gastric nerves, occasioned by an interval of inactivity, during which the vital powers may be supposed to accumulate,"‡ it appears to me that we are venturing upon unexplored grounds, of which we know

‡ Paris On Diet, p. 55.

but little. We are not accustomed to call those painful nervous sensations to which the system is sometimes subject, states of high nervous energy. Are they not rather states of nervous debility? or, at any rate, irregular and unhealthy motions?

That the introduction of narcotics into the stomach should destroy the appetite, proves only that they have the same effect on that organ as they have on other parts of the body; they paralyse the nerves, and render them incapable of being the media of communication to their common centre.

Many other causes have been assigned for this sensation, equally wide, probably, of the true one. It has been attributed to the "foresight of the vital principle," a phrase that means any thing, every thing, or nothing, according to the construction which each one may put upon it. Such explanations conduce nothing to the promotion of science. They are mere sounds and words, which ingeniously convey a tacit acknowledgement of their author's ignorance.

Again, the mechanical action of the liver upon the diaphragm, has been accused of producing the sensation of hunger. Some proof, more than mere assertion, is necessary to convince honest enquirers that so remote a cause should produce such effects on the stomach, the immediate seat of the feeling. Of the same nature, is the opinion of the fatigue of the contracted fibres of the stomach, or of compression of the nerves of that organ, &c. &c.

Magendie, convinced that all the theories on this subject were unsatisfactory, comes to the following comprehensive conclusion: that "Hunger is produced like all other internal sensations, by the action of the nervous system, and it has no other seat than in this system itself, and no other causes than the general laws of organization."† I cannot perceive that such explanations bring the mind to any satisfactory understanding of the subject. In such broad propositions, it is difficult to ascertain the exact meaning. If the design is to convey the impression that hunger has no "local habitation;" that it is an impression, affecting all the nerves of the system in the same manner; then the sensation would be as likely to be referred to one organ as another. It is true, that without nervous communication there would be no sensation at all. This applies as well to other parts as to the stomach. The nerves are the media of communication from the sensible parts to the centre of perceptions. They warn the encephalon not only of the injuries, but of the wants of the tissues. We are accustomed to refer local sensations and irritations to the parts *apparently* affected—desire for urination and defecation, to the bladder and rectum; for liquids, to dryness of the mouth and fauces: and we account, in like manner, for other physiological and pathological sensations. When we can arrive at the exact interpretation of an author, who says that hunger has "no other causes than the general laws of organization," it will then be time to give reasons for an assent to or dissent from the proposition.

† Summary of Physiology, p. 196.

This subject is, unquestionably, involved in considerable doubt and obscurity, and will not, it is to be apprehended, admit of a very speedy elucidation. The Author of Nature is perfect in all His works; and although we may not understand all the operations of His hands, we are compelled to acknowledge their wisdom, propriety and beauty. Man would be miserable and wretched indeed, if he depended solely on his own discretion and judgment to decide upon the quantity and quality of aliment necessary to supply the wastes, and administer to the growth, of the system. This paucity of judgment and discretion is, however, more than compensated by an irresitible sensation, which indicates the proper time for the reception of food. The immediate cause of this sensation, as we have seen, has not as yet received a very satisfactory explanation, and perhaps will not admit of one. But, although confessedly obscure, we are not denied the privilege of patient investigation, and persevering search after truth. Knowledge is progressive, as well in this as every other science; and every new discovery, and every rational hypothesis, are additions to the general stock. Persuaded of the truth of these general propositions, and anxious mainly to elicit investigation on the subject, I submit the following *Theory of Hunger*, believing it to be as reasonable, to say the least, as any that has been propagated.

My impression is, that the sensation of hunger is produced by a *distention* of the gastric vessels, or that apparatus, whether vascular or glandular, which secretes the gastric juice; and is believed to be the effect of repletion by this fluid.

One reason, among others, for this belief, is the established fact, that the internal sensations referred to different organs, as has been previously alluded to, are caused by some modified action or condition of the parts in the tissues of the organ itself. The modification in the parts to which the sense of hunger is invariably referred, I conceive to be a distention, by the gastric juice, of a particular set of vessels or glands, constituting, in part, the erectile tissue of the villous coat of the stomach. The sensation varies according to the different degrees or states of distention, from the simplest desire to the most painful sense of hunger; and is allayed or increased in proportion to the application, or refusal, of alimentary stimulus to the excretory vessels. The greater the distention of the vessels, the more acute will be the pain: hence, the difference between a short and protracted fast. Appetite and hunger belong to the same class of sensations; they differ only in degree. In this they are like all other sensations. A little increased circulation in the vessels of the brain produces peculiarly vivid, but not absolutely unpleasant feelings, and gives force and energy to the mental volitions: carried further, it produces most painful sensations. It is unnecessary to cite further examples. Indeed, it does not need arguments to prove what is the subject of every day's observation. It is well known that the pain from acute inflammation is produced by distention of the blood vessels. Let any one, who is disposed to try the effect of vascular distention, place a ligature around the finger or arm, sufficiently tight to retard the returning blood, and the truth will be sufficiently obvious.

It is, therefore, inferred from the pain, (and no one, it is believed, will deny that *hunger* is a painful sensation, whatever may be his opinion of *appetite*) that vessels of some kind are distended; and it is demonstrated, I think, in some of the following experiments, that these are the gastric vessels. On applying aliment to the internal coat of the stomach, which, in health, is merely lubricated with mucus, innumerable minute papillæ, the orifices, undoubtedly, of the gastric vessels, immediately throw out a quantity of the fluid, which mixes with the food. This effect is too sudden, and the secretion too copious, to be accounted for on the ordinary principles and laws of secreting mucous surfaces. The quiescence and relief from the unpleasant sensation, which are experienced as soon as the vessels are emptied, are, I think, additional proofs of my opinion. It is certain, that at the introduction of every meal, or on the application of alimentary stimulus to the internal coat of the stomach, a very large secretion of a fluid, which has repeatedly been ascertained to be an alimentary solvent, immediately takes place; and that when the stomach is destitute of food or some other irritating substance, no such secretion can be found in it. And it is more than probable—it, in fact, almost amounts to demonstration, that a large quantity of this fluid must be contained in appropriate vessels, during a fast, ready to obey the call of aliment. I would not be understood to say that the whole quantity necessary for an ordinary meal is eliminated from the blood, previous to the commencement of alimentation; but that enough is contained in the gastric vessels to produce the sensation of pain or hunger.

If it be objected to this theory, that the vessels would become ruptured, or empty themselves into the cavity of the stomach, during a long fast, I reply, that this apparatus is probably constituted like many of the other organs of the system, and permits the absorption of its secretions by the lymphatic or other absorbent vessels. The male semen is constantly being secreted, and deposited in its proper seminal vessels, ready to be ejected during the venerial orgasm; and yet how many men live for years, or perhaps for a whole life, who have no intercourse with the other sex. What becomes of the semen under these circumstances? Taken up, unquestionably, by the absorbing vessels, as the gastric juice of the stomach is.

I offer this theory for consideration, persuaded that the public will allow it such weight as it may have a right to claim: more than this, I have no wish to ask.

Thirst.—This sensation is felt in the mouth and fauces.. Like hunger, it is a kind provision of nature, designed to remind men and animals of the necessity, not of replenishing the wasting solids of the system, but of diluting the fluids that are carrying on these processes. Although Magendie has attempted to put a stop to all inquiries on this subject, in the remark, that "Thirst is an internal sensation, an instinctive sentiment;" "the result of organization, and does not admit of any explanation;" I apprehend a remote cause of this sensation may be found in the viscidity of the blood, which requires a liquid to render it more fluid, and more susceptible of introduction into the capillaries and secreting surfaces. The proximate cause may exist in an irritation, a kind of sub-inflammation of the mucous membranes of the mouth and fauces, the effect of the viscid state of the blood, and consequently impervious state of the secretory vessels of these membranes. The sensation of dryness, or thirst, is supposed to be the effect of evaporation, the mouth and throat being constantly exposed to the atmosphere. When there is sufficient fluidity of the blood, the secretion is so much more copious than the evaporation, that a constant moisture is preserved. The sensation of thirst resides in the tissues; and it is no more "an instinctive sentiment" than any other sensation of the economy. To say that it is the "result of organization," gives no explanation, amounts to nothing, and is certainly, to say the least, a very unsatisfactory way of disposing of the question.

SECTION III.

Of Satisfaction and Satiety.

IN the present state of civilized society, with the provocatives of the culinary art, and the incentives of high seasoned food, brandy and wines, the temptations to excess in the indulgences of the table are rather too strong to be resisted by poor human nature. It is not less the duty, however, of the watchmen on the walls to warn the city of its danger, however it may regard the premonition. Let them at least clear their own skirts from the stain of unfaithfulness, whatever may be the result.

There is no subject of dietetic economy about which people err so much, as that which relates to *quantity*. The medical profession, too, have been accessory to this error, in giving directions to dyspeptics to eat until a sense of satiety is felt. Now, this feeling, so essential to be rightly understood, never supervenes until the invalid has eaten too much, if he have an appetite, which seldom fails him. Those, even, who are not otherwise predisposed to the complaint, frequently induce a diseased state of the digestive organs by too free indulgence of the appetite. Of this fact the medical profession are, generally, not sufficiently aware. Those who lead sedentary lives, and whose circumstances will permit of what is called free living, are peculiarly obnoxious to these complaints. But by paying particular attention to their sensations during the ingestion of their meals, these complaints may be avoided. There appears to be a sense of perfect intelligence conveyed from the stomach to the encephalic centre, which, in health, invariably dictates what quantity of aliment (responding to the sense of hunger, and its due satisfaction,) is naturally required for the purposes of life; and which, if noticed, and properly attended to, would prove the most salutary monitor of health, and effectual preventive of, and restorative from, disease. It is not the sense of *satiety*, for this is beyond the point of *healthful* indulgence, and is nature's earliest indication of an *abuse* and *overburthen* of her powers to replenish the system. It occurs immediately previous to this, and may be known by the pleasurable sensation of *perfect satisfaction, ease and quiescence of body and mind.* It is when the stomach says *enough*, and is distinguished from satiety by the difference of the sensations—the former feeling *enough*—the latter, *too much.* The first is produced by the timely reception into the stomach of proper aliment, in exact proportion to the requirements of nature, for the perfect digestion of which, a definite quantity of gastric juice is furnished by the proper gastric apparatus. But to effect this most agreeable of all sensations and conditions—the real Elysian satisfaction of the *reasonable* Epicure—timely attention must be paid to the preliminary processes, such as thorough mastication, and moderate or slow deglutition. These are indispensable to the due and natural supply of the stomach, at the stated periods of alimentation; for if food be swallowed too fast, and pass into the stomach imperfectly masticated, too much is received in a short time, and in too imperfect a state of preparation, to be disposed of by the gastric juice.

The quantity of gastric juice, either contained in its proper vessels, or in a state of preparation in the circulating fluids, is believed to be in exact proportion to the proper quantity of aliment required for the due supply of the system. If a more than ordinary quantity of food be taken, a part of it will be left undissolved in the stomach, and produce the usual unpleasant symptoms of indigestion. But if the ingestion of a large quantity be in proportion to the calls of nature, which sometimes happens after an unusual abstinence, it is probable that more than the usual supply of gastric juice is furnished; in which case the apparent excess is in exact ratio to the requirements of the economy; and never fails to produce a sense of quiescent gratification, and healthful enjoyment. A great deal depends upon habit, in this respect. Our western Indians, who frequently undergo long abstinences from food, eat enormous quantities, when they can procure it, with impunity.

Satiety is produced by tendering too much at once for the wants of the economy; more than the gastric juice is able to dispose of at the time; distending the muscular fibres beyond that point so admirably fixed, by the invariable and universal laws of the animal system, for agreeable sensations; disturbing the peculiarly pleasurable, undulatory motions of the rugæ of the stomach, in their operations of forming chyme; and *perhaps*, interrupting, if not diminishing, the secretion of the gastric juice. The redundant aliment, incapable of being dissolved, for want of sufficient gastric juice, remains, and becomes a source of irritation, and renders imperfect the chymification of that which would otherwise have been completed. Hence the sense of weight, and disagreeable fulness, attendant on an unusually hearty meal; the subsequent derangement of the digestive functions, and consequent acidities and vitiated contents of the prima viæ, from acetic fermentation in the stomach, and imperfect formation of chyle in the intestines.

SECTION IV

Of Mastication, Insalivation and Deglutition.

THESE are the preliminary steps in the process of digestion. The comparative importance of these processes has been elevated or depressed, according to the preponderance which each of them may have received from the opinions of the different physiologists who have made them subjects of observation. As man and animals are constituted, they are all absolutely necessary to the digestion of food. But in an abstract point of view, disconnected as a mean of introducing ingestæ into the stomach, I believe I hazard nothing in saying that they may be considered as perfectly non-essential to chymification. If the *materia ulimentaria* could be introduced into the stomach in a finely divided state, the operations of mastication, insalivation and deglutition, would not be necessary. Aliment is as well digested and assimilated, and allays the sensation of hunger as perfectly, when introduced directly into the stomach, in a proper state of division, as when the previous steps have been taken, as may be seen by some of the following experiments. If particular importance is to be attributed to any of these previous steps, it is certainly due to mastication; though an undue importance has, of late, been given to the action of the *saliva*. Professor Jackson, of Philadelphia, who has lately published a physiological work on the "Structure and Functions of the Animal Organism," has elevated saliva to a rank in the process of digestion, seldom before claimed for it. He considers it the principal solvent, or macerating agent, of alimentary matter. He is sustained in this opinion by Montegre and others. Even Magendie is inclined to favour this belief.

It is remarked by Paris, (On Diet, p. 37,) that the introduction of saliva into the stomach is "obviously essential to a healthy digestion." That it is generally introduced into the stomach with the food is very obvious; the nature of its action is not so clear. In most of the experiments that follow, artificial digestion was performed without the admixture of saliva. Chyme formed in this way, exhibited the same sensible appearances, and was affected by re-agents in the same way, as that which was formed from food which had been previously masticated, mixed with the saliva, and swallowed. It would seem, from two or three of the experiments on artificial digestion, which were instituted for the purpose of comparison, that the mixture of saliva with the gastric juice rather retarded its solvent action. But I do not wish to deny the utility of the saliva. It is certainly important as a preliminary to digestion. Its legitimate and only use, in my opinion, is to lubricate the food, and to facilitate the passage of the bolus through the organs of deglutition. In this point of view, it is essential,

Dry food cannot be swallowed until it receives an admixture of a fluid, whether it be saliva or some other liquid, is not, I conceive, a matter of much importance. Any one, disposed to try the experiment, may satisfy himself of this fact, by attempting to swallow a mouthful of dry cracker, meal or magnesia. He will find it impossible to make the organs of deglutition act till a quantity of fluid is mixed with it. Water will answer the purpose, nearly as well as saliva; though the mucous properties of this secretion may give it a slight preference.

Pathology is not, in my opinion, much indebted to Ruysch, who attributed the loss of appetite to the waste of saliva in a person who was afflicted with a fistula in one of the salivary ducts; nor to the opinion advanced by others, that the constant spitting of maniacal patients, induces loss of appetite. The truth is, that in both cases, the effects are attributed to the wrong causes. There is no difficulty in believing that a foul ulcer in the mouth would be liable to produce nausea and want of appetite; nor that maniacal patients are generally, if not always, affected with diseased organs of digestion. I have known many persons to spit freely and constantly, whose appetites and digestion were perfect. Those who smoke tobacco are constantly discharging large quantities of saliva; and yet I am not aware that dyspepsia is more common with them than with others.

I entirely dissent from the opinion advanced by the author above referred to, (Paris,) that "Insalivation is as essential as mastication." The use of mastication is to separate the food into small particles, so that the solvent of the stomach may be applied to a greater extent of surface. There is no mystery about this. Every body knows that the smaller the particles of matter that are submitted to the action of a chemical agent, the more vigorously the agent will act upon them, and the sooner they will be dissolved, or decomposed.‡ Mastication is absolutely necessary to healthy digestion. If aliment, in large masses, be introduced into the stomach, though the gastric juice may act upon its surface, chymification will proceed so slowly, that other changes will be likely to commence in its substance before it will become completely dissolved. Besides, the stomach will not retain undigested masses for a long time, without suffering great disturbance. It is governed by certain laws with respect to aliment. After food has been retained for a certain length of time undigested, say from five to ten hours, according to the healthy or diseased state of that organ, or the quantity received into it, it is either rejected by vomiting, or is permitted to pass into the duodenum and lower bowels, where its presence almost invariably produces colic, flatulence, &c. When the stomach is unusually debilitated, food, however, is frequently retained for twenty-four hours or more, and is sometimes the cause of most distressing symptoms, producing, particularly in children, convulsions and death. I therefore consider mastication as one of the most important preliminary steps in the process of digestion.

‡ In using the word *solvent* or *solution*, in reference to the gastric juice, I wish to be understood to mean a chemical action, analogous to that of the action of mineral acids on the metals; not like the solution of sugar or salt in water.

With respect to deglutition, I shall make but a few remarks. It is important for the preservation of health, that this process should be effected slowly. If food be swallowed rapidly, more will generally be taken into the stomach before the sensation of hunger is allayed, than can be digested with ease. If due attention be paid to the previous step of thorough mastication, we shall not be so likely to err in this later one.

Swallowing very rapidly, produces irregular contractions of the muscular fibres of the œsophagus and stomach; disturbs the vermicular motions of the rugæ, and interrupts the uniform action of the gastric apparatus.

The stomach is not designed to receive more food than can be duly mixed with the gastric solvent, already in its proper vessels, or in a state of preparation in the blood vessels. Perfect harmony of action must exist throughout the whole apparatus, or derangement of healthy action will ensue.

The stomach will not admit of the introduction of food, even of a liquid kind, through the aperture, at a rapid rate. If a few spoonfuls of soup, or other liquid diet, be put in with a spoon or funnel, the rugæ gently close upon it, and gradually diffuse it through the gastric cavity, entirely excluding more during this action. When a relaxation takes place, another quantity will be received in the same manner.

If the valvular portion of the stomach be depressed, and solid food be introduced, either in entire pieces, or finely divided quantities, the same gentle contraction, or grasping motion, takes place, and continnes for fifty or eighty seconds; and will not allow of the introduction of another quantity until the above time has elapsed; when the valve may again be depressed, and more food be put in. Food and drinks will be received through the aperture no faster, even when the stomach is entirely empty, than they are ordinarily received through the œsophagus.

When the subject of these experiments is so placed that the cardia can be seen, and he be allowed to swallow a mouthful of food, the same contraction of the stomach, and closing upon the bolus, is invariably observed to take place at the œsophageal ring.

SECTION V.

Of Digestion by the Gastric Juice.

CHYMIFICATION is effected in the stomach. It is the first stage, proper, of the conversion of aliment into blood; though in the ordinary course of proceeding, as animals are constituted, some previous steps are necessary. After the aliment has been received into the stomach, it is subjected to certain evolutions, or motions, propagated by the muscular fibres of that organ; and is acted upon through the agency of some principle, which changes it from a heterogeneous mixture of the various kinds of diet, submitted to its action, to an uniform, homogeneous semi-fluid, possessing properties distinct from the elements of which it was composed. The length of time consumed in the operation is various. It depends upon the quantity or quality of the ingestæ, the healthy or diseased state of the stomach, &c. In the various experiments which I have made, the medium time may be calculated at about three and a half hours.

It has been suggested by many physiologists, and positively asserted by some, that there is considerable increase of the temperature of the stomach during the digestion of a meal. But from the result of a great number of experiments and examinations, made with a view of ascertaining the truth of this opinion, in the empty and full state of the organ, and during different stages of chymification, I am convinced that there is no alteration of temperature, unless some other circumstance should produce it. Active exercise always elevates the temperature of the stomach, whether fasting or full, about one and a half degrees.

With respect to the agent of chymification, that principle of life which converts the crude aliment into chyme, and renders it fit for the action of the hepatic and pancreatic fluids, and final assimilation and conversion into the fluids, and the various tissues of the animal organism—no part of physiology has, perhaps, so much engaged the attention of mankind, and exercised the ingenuity of physiologists. It has been a fruitful source of theoretical speculation, from the father of medicine down to the present age. It would be a waste of time to attempt to refute the doctrines of the older writers on this subject. Suffice it to say, that the theories of *Concoction*, *Putrefaction*, *Trituration*, *Fermentation* and *Maceration*, have been prostrated in the dust before the lights of science, and the deductions of experiment. It was reserved for SPALLANZANI to overthrow all these unfounded hypotheses, and to erect upon their ruins, a theory which will stand the test of scientific examination and experiment. He established a theory of CHEMICAL SOLUTION, and taught that chymification was owing to the solvent action of a fluid, secreted by the stomach, and operating as a true menstruum of alimentary substances. To this fluid he gave the name of GASTRIC JUICE. It does not come within the scope of this work to give a detail of the experiments and reasoning which wrought conviction in the mind of this great man. It is only necessary to say that it was the result of patient and persevering experiment and research.

The truth of SPALLANZANI'S theory has been sustained, so far as relates to the most important part, the existence of a chemical solvent, by all who have made fair examinations and experiments on the subject. The experiments of TIEDEMANN and GMELIN, of LEURET and LASSAIGNE, confirm the same theory.

By far the most respectable and intelligent physiologists have now settled down in the belief that chymification is effected in the stomach, by a peculiar and specific solvent, secreted in that organ, called, after SPALLANZANI, the Gastric Juice. From the difficulty, however, of obtaining and submitting such fluid to the test of experiment, and the diversity of results in the examination of such as has been obtained, much indefiniteness is experienced on this subject. The presence of an active solvent is rather an admission on their part—a conclusion from the effect to the cause. BROUSSAIS, speaking on this subject, says: "It remains for us to know whether the portion of mucous membrane, belonging to the stomach, contains secretory organs, the office of which is to furnish a fluid, fit to produce the assimilation of nutritive substances." And, again, speaking of the gastric juice, "The question is as yet *undecided*, though, if we are to judge by analogy, we shall observe that many animals are furnished with gastric glands, supplying a digestive liquid." This author *admits* the presence of a solvent fluid in the stomach, without, however, attempting to explain its specific effects, or mode of operation; for he says, in another place, "We have expressed our opinion on this subject; but whether the gastric fluids possess an assimilating property, which, for ourselves, we admit, without pretending to demonstrate its actual presence," &c.

Richerand, Bostock, and nearly all the authors of modern date, teach the doctrine of digestion by the gastric juice, without, however, pretending to explain its exact mode of operation. Professor Dunglison, whose work on "Human Physiology," taken as a whole, is, perhaps, the most comprehensive, arrives at the same conclusion. He says, "We have too many evidences in favour of the chemical action of some secretion from the stomach during digestion, to permit us to doubt for a moment of the fact." And, again—"From all these facts, then, we are justified in concluding that the food in the stomach is subjected to the action of a secretion, which alters its properties, and is the principal agent of converting it into chyme."

I have referred to these learned authors with the view of showing the exact state of the science on this subject.

Though the theory of chymification by the gastric juice, has become almost universal with physiologists, and the medical profession in general, still there are some, even of very modern date, who, with all the lights of science and experiment, from aversion to the slow and tedious processes by which truths are attained, or, perhaps, from the ambition of becoming the discoverers of some new and extraordiny process, or the projectors of some fanciful theory, deny the power of the gastric juice, or even the existence of such a fluid; and set at naught the experiments, observations and opinions of the ablest physiologists, and most experienced writers on this subject.

That chymification is effected by the *solvent* action of the gastric juice, aided by the *motions* of the stomach, and the natural *warmth* of the system, not a doubt can remain in the mind of any candid person, who has had an opportunity to observe its effects on alimentary substances, or who has the liberality to credit the opinions of those who have had such opportunities.

It has been objected to this hypothesis, that the *sensible* properties of the gastric juice contradict the opinion of its active *solvent* effect. But we should recollect that many things which make very little impression on our external senses, produce most astonishing effects in other situations. The air which we breathe, by which we are surrounded, and which, to our external senses, is almost inappreciable, is one of the most powerful and destructive agents in nature—one portion of which is capable of combining with all grades of matter, either slowly and imperceptibly, as in the gradual decay of all substances, or rapidly, as in the combustion of wood, or even the hardest metals—and which by means, inexplicable to us, sustains in life and being the whole of animated nature.

The gastric juice has been submitted to chemical examination and analysis, with various results. Perhaps in the present state of the science of chemistry it will not be practicable to ascertain its exact chemical character. The parcels heretofore submitted to analysis, have been very impure; but the result of even these partial examinations, has been to show that this fluid contains a portion of free muriatic acid, combined with the acetic, and some salts. In the winter of 1832-3, I submitted a quantity of gastric juice, with no other admixture, except a small proportion of the mucus of the stomach, to Professor DUNGLISON, for examination, who, with the assistance of the professor of chemistry of the Virginia University, effected the following analysis, and was kind enough to communicate the result to me by letter.

"UNIVERSITY OF VIRGINIA,
Feb. 6th, 1833.

"MY DEAR SIR,

"Since I last wrote you, my friend and colleague, Professor Emmett, and myself, have examined the bottle of gastric fluid which I brought with me from Washington, and we have found it to contain free *Muriatic* and *Acetic* acid, *Phosphates* and *Muriates*, with bases of *Potassa*, *Soda*, *Magnesia* and *Lime*, and an *animal matter, soluble in cold water, but insoluble in hot.* We were satisfied, you recollect, in Washington, that free muriatic acid was present, but I had no conception it existed to the amount met with in our experiments here. We distilled the gastric fluid, when the free acid passed over; the salts and animal matter remaining in the retort. The quantity of Chloride of Silver thrown down on the addition of the Nitrate of Silver, was astonishing."

I had been long convinced of the existence of free muriatic acid in the gastric fluids. Indeed, it is quite obvious to the sense of taste; and most chemists agree in this, however they may be at variance with respect to the other constituents. The analysis of Professors DUNGLISON and EMMETT is certainly as satisfactory as any that has as yet been made. It is a question, too, whether gastric juice, in so great a state of purity, has ever before been submitted to chemical analysis.

It is to be hoped that no one will be so disingenuous as to attribute to Professor DUNGLISON the design of finding the existence of certain chemical agents in the gastric juice, with the view of propping the theory of the chemical action of this fluid, which he has maintained in his work on "Human Physiology;"—or, in other words, to say, that he had determined to find certain results; and that he had accordingly found them. Those who are acquainted with him, know that his candour and fairness are above the reach of suspicion; and that he would be equally willing to retract a false opinion as to maintain a correct one. Another quantity was sent to him for further analysis; but I regret that no report has yet been received from him.

In April of the present year, (1833,) a parcel was submitted to BENJAMIN SILLIMAN, M. D. Professor of Chemistry in Yale College. Professional engagements prevented his examination of the fluid until the 2d of August, when he sent me the following result:

" *Examination of the Gastric Fluid, Aug.* 2, 1833.

" 1. The Fluid, after being kept in a closely corked vial, more than three months, from April to August, and most of the time in a cellar, remained unaltered, except the formation of a pellicle upon the surface, slightly discoloured by red spots. A second pellicle appeared after the precipitation of the first. It was thicker, and more discoloured with dark red spots, like venous blood.

" 2. The Fluid was cloudy, like a solution of gum arabic; but on filtering, it became perfectly clear, and of a slight, straw yellow tinge.

" 3. The pellicles, which had the appearance of inspissated mucus, after being separated from the Fluid, became, after exposure to the air, throughout of a brownish red colour, resembling the inner portion of a mass of coagulated blood. This change seemed to result from a sudden oxygenation.

" 4. The Fluid exhaled a slight odour—not disagreeable—rather aromatic—and very similar to that which it at first exhaled; but not so strong. It was then rather disagreeable.

" 5. Taste, feebly saline—not disagreeable.

" 6. Test papers of litmus, alkanet, and purple cabbage, were decidedly reddened. Turmeric paper underwent no change: but when previously browned by an alkali, (ammonia) the gastric fluid restored the yellow colour.

" 7. Nitrate of Silver gave a dense white precipitate, which, after standing five minutes in the sun's light, turned to a dark, brownish black; thus indicating Muriatic Acid. Mur. and Nit. Barytes gave a slight opalescence, indicating a trace of sulphuric acid; not improbably, there was also some phosphoric acid.

"8. Specific gravity—when taken in a small, thin glass tube, containing 201 grs. of distilled water—when filled with the gastric fluid, its weight was increased 1 gr.—weight of the gastric liquor, therefore, 202 grs. The specific gravity is, therefore, about 1.005. But little solid matter in solution."

At the instance of Professor SILLIMAN, I committed to the care of Mr. GAHN, Consul of his Swedish Majesty in New-York, a bottle, containing one pint, of gastric juice, to be transmitted by him to Professor BERZELIUS, of Stockholm, one of the most eminent chemists of the age, with a request that he would favour me with an analysis. Some unavoidable delay was experienced in forwarding the bottle; and no returns have yet been received. It is hoped, however, that they will arrive in time to be attached in an appendix to this volume.

The following results have been obtained from partial examinations and analyses of the gastric juice, or rather, in most instances, of the *mixed fluids* of the stomach.

SPALLANZANI, in 1793, after many experiments, declared the gastric juice to be entirely *neutral*, a *solvent* for alimentary matter, *within* and *without* the stomach—that it did not *putrefy* at the ordinary temperature of the stomach; but preserved animal matters from putrefaction, and dissolved them, with the aid of heat.

SCOPOLI found in the gastric juice of the rook, water, gelatine, a saponaceous matter, muriate of am monia and phosphate of lime.

CARMINITI, in 1795, found it in carnivorous animals,

salt and bitter, and frequently acid when they had eaten, but not so when fasting.*

VIRIDET, WERNER, HUNTER and others, found the gastric juice acid.

MM. MARQUART and VAUQUELIN found albumen and free phosphoric acid in it.

TIEDEMANN and GMELIN found it to contain, on analysis, muriatic and acetic acid; mucus; very little or no albumen; salivary matter; osmazome; muriate and sulphate of soda. In the ashes, carbonate, phosphate and sulphate of lime, and chloride of calcium. Principally from carnivorous animals.

LEURET and LASSAIGNE, in a hundred parts, found, water, ninety-eight, lactic acid, muriate of ammonia, muriate of soda, animal matter soluble in water, mucus, and phosphate of lime, two parts.

MONTEGRE, (1812) *who could vomit at will*,‡ and who analyzed the fluid so obtained, declared it not to be *acid*—not a *solvent*—not *slow to putrefy*—so much like *saliva*, that he regards it saliva swallowed.

PROUT, 1824, declares the gastric juice to be really *acid*—does not contain an organic acid, but free, hydrochloric, or muriatic acid.

These opinions are certainly discordant. The majority of evidence, however, is in favour of the existence of pretty active chemical agents in the gastric fluids—perhaps not sufficient, in comparison with

* Probably because the fluid found in the stomach when fasting, was not gastric juice.

‡ See remarks near the close of this section on Montegre's experiments.

the ordinary operations of chemistry, to account for the digestion, or solution of aliment.

The discrepance of results in the reports of those who have had opportunities of examining the process of, and have made experiments on, *artificial digestion*, by the gastric juice, as well as in the chemical examination of this fluid, has been owing more to the difficulty of obtaining it pure, in sufficient quantity, and under proper circumstances, than to any real difference in its effects. Under the circumstances in which the following experiments were made, I flatter myself that these difficulties have been obviated; and if the inferences are incorrect, the blame must be attached to the experimenter. He can only say, that the experiments were made in good faith, and with a view to elicit facts.

I think I am warranted, from the result of all the experiments, in saying, that the gastric juice, so far from being "inert as water," as some authors assert, is the most general solvent in nature, of alimentary matter—even the hardest bone cannot withstand its action. It is capable, *even out of the stomach*, of effecting perfect digestion, with the aid of due and uniform degrees of heat, (100° Fahrenheit,) and gentle agitation, as will be seen in the following experiments.

The fact that alimentary matter is *transformed*, in the stomach, into chyme, is now pretty generally conceded. The peculiar process by which the change is effected, has been, by many, considered a problem in physiology. Without pretending to explain the exact *modus operandi* of the gastric fluid, yet I am impelled by the weight of evidence, afforded by the experiments, deductions and opinions of the ablest physiologists, but more by direct experiment, to conclude that the change effected by it on aliment is *purely chemical.* We must, I think, regard this fluid as a chemical agent, and its operation as a chemical action. It is certainly every way analogous to it; and I can see no more objection to accounting for the change effected on the food, on the supposition of a chemical process, than I do in accounting for the various and diversified modifications of matter, which are operated on in the same way. The decay of the dead body is a chemical operation, separating it into its elementary principles—and why not the solution of aliment in the stomach, and its ultimate assimilation into fibrine, gelatine and albumen? Matter, in a natural sense, is indestructible. It may be differently combined; and these combinations are chemical changes. It is well known that all organic bodies are composed of very few simple principles, or substances, modified by excess or diminution of some of their constituents.

The gastric juice appears to be secreted from numberless vessels, distinct and separate from the mucous follicles. These vessels, when examined with a microscope, appear in the shape of small lucid points, or very fine papillæ, situated in the interstices of the follicles. They discharge their fluid only when solicited to do so, by the presence of aliment, or by mechanical irritation.

Pure gastric juice, when taken directly out of the stomach of a healthy adult, unmixed with any other fluid, save a portion of the mucus of the stomach, with which it is most commonly, and perhaps always combined, is a clear, transparent fluid; inodorous; a little saltish; and very perceptibly acid. Its taste, when applied to the tongue, is similar to thin mucilaginous water, slightly acidulated with muriatic acid. It is readily diffusible in water, wine or spirits; slightly effervesces with alkalis; and is an effectual solvent of the *materia alimentaria.* It possesses the property of coagulating albumen, in an eminent degree; is powerfully antiseptic, checking the putrefaction of meat; and effectually restorative of healthy action, when applied to old, fœtid sores, and foul, ulcerating surfaces.

Saliva and mucus are sometimes abundantly mixed with the gastric juice. The mucus may be separated, by filtering the mixture through fine linen or muslin cambric. The gastric juice, and part of the saliva will pass through, while the mucus, and spumous or frothy part of the saliva, remains on the filter. When not separated by the filter, the mucus gives a ropiness to the fluid, that does not belong to the gastric juice, and soon falls to the bottom, in loose, white flocculi. Saliva imparts to the gastric juice, an azure tinge, and frothy appearance; and, when in large proportion, renders it fœtid in a few days; whereas the *pure* gastric juice will keep for many months, without becoming fœtid.

The gastric juice does not accumulate in the cavity of the stomach, until alimentary matter be received, and excite its vessels to discharge their contents, for the immediate purpose of digestion. It then begins to exude from its proper vessels, and increases in proportion to the quantity of aliment *naturally* required, and received. A definite proportion of aliment, only, can be perfectly digested in a given quantity of the fluid. From experiments on artificial digestion, it appears that the proportion of juice to the ingestæ, is greater than is generally supposed. Its action on food is indicative of its chemical character. Like other chemical agents, it *decomposes*, or *dissolves*, and combines with, a fixed and definite quantity of matter, when its action ceases. When the juice becomes *saturated*, it refuses to dissolve more; and, if an excess of food have been taken, the residue remains in the stomach, or passes into the bowels, in a crude state, and frequently becomes a source of nervous irritation, pain and disease, for a long time; or until the *vis medicatrix naturæ* restores the vessels of this viscus to their natural and healthy actions—either with or without the aid of medicine.

Such are the appearance and properties of the gastric juice; though it is not always to be obtained pure. It varies with the changing condition of the stomach. These variations, however, depend upon the admixture of other fluids, such as saliva, water, mucus, and sometimes bile, and, perhaps, pancreatic juice. The special solvent itself—the *gastric juice*—is, probably, invariably the same substance. Derangement of the digestive organs, slight febrile excitement, fright, or any sudden affection of the passions, cause material alterations in its appearance. Overburthening the stomach produces acidity and rancidity in this organ, and retards the solvent action of the gastric juice. General febrile irritation seems entirely to suspend its secretion into the gastric cavity; and renders the villous coat dry, red and irritable. Under such circumstances, it will not respond to the call of alimentary stimulus. Fear and anger check its secretion, also:—the latter causes an influx of bile into the stomach, which impairs its solvent properties.

When food is received into the stomach, the gastric vessels are excited by its stimulus to discharge their contents, when chymification commences. It has been a favourite opinion of authors, that food, after it has been received into the stomach, should "remain there a short period before it undergoes any change;"* the common estimate is one hour. But this is an erroneous conclusion, arising from inaccuracy of observation. Why should it remain there, unchanged? It has been received into the organ which is to effect an important change upon it—the gastric juice is ready to commence its work of solution soon after the first mouthful is swallowed; and, certainly, if we admit that the gastric juice performs the office of a chemical agent, which most physiologists allow, it is contrary to all our notions of chemical action, to allow it one moment to rest. It must commence its operation immediately. That it does so, is distinctly manifested by close observation of its action on food, in the healthy stomach.

But Paris is not alone in this opinion. It appears to have been a favourite doctrine; and has been regularly handed down, from one physiologist to another, as a sort of *heir loom* to the profession. The suc-

* Paris On Diet, p. 39.

cessors in the physiological sciences seem to have been compelled to receive it with the legacy of their predecessors, without any doubt of its legitimacy; when, with a little rational examination of the subject, it would have been found a fair subject of rejection. It will be seen, by the following experiments, that it has not the slightest foundation in truth; and to them I refer the reader.

It has been said, that when one meal follows another in quick succession—or, in other words, when a subsequent meal is taken before the previous one is digested—that it *some how* disturbs the process of digestion. This is generally true; and it allows of a definite solution. It is because more is received into the stomach, in the aggregate, than the gastric juice can dissolve. And this disturbance will result, as well when too much food has been taken at once, as when too much has been received in rapid succession. But if the quantity be moderate, no ill effect will ensue. Many children are in the habit of eating as often as once an hour through the day, in small quantities, without experiencing any bad consequences. Cooks are, also, accustomed to the practice of constantly tasting of the various articles of food which they are preparing for the table; and yet I am not aware that they suffer any inconvenience from the habit. From these, and other facts, as well as from direct experiment, I think it is perfectly apparent that digestion must progress as well before as after the expiration of an hour. If, as has been suggested, the ingestion of food, in addition to the delay to itself, retards or stops the chymification of that which had been previously received, aliment, as it relates to those children who eat hourly, would be constantly accumulating; and there would remain in the stomach at night the whole quantity taken through the day: a supposition not to be credited, even by those disposed to make the most of a favourite opinion or doctrine.

Doctor Wilson Philip, in his Treatise on Indigestion," says, "the layer of food lying next to the surface of the stomach, is first digested, and in proportion as this undergoes the proper change, and is moved by the muscular action of the stomach, that next in turn succeeds, to undergo the same change." That chymification commences on the surface of the food, I have no doubt; but I apprehend this to be the case as it respects each individual portion, and not the whole mass. I have frequently taken out portions from the stomach, a few minutes after they had been received into that organ, when they appeared to have received a full supply of gastric juice for perfect digestion, when submitted to the artificial mode. When a due and moderate supply of food has been received, it is probable that the whole quantity of gastric juice for its complete solution, is secreted, and mixed with it, in a short time. When an unusually full meal has been eaten, the necessary quantity for its complete solution, is not so readily supplied. If a tenacious mass of food be used, the external portion of the whole quantity is first acted on, digested, and succeeding portions presented, &c. There is no ground for the opinion inferred, that the gastric juice never leaves the parietes of the stomach, except as it chymifies food. It is a thin fluid, and is governed by the same laws that other thin fluids are. From numerous examinations of the stomach, I feel warranted in saying, at least in the human subject, that there is a perfect admixture of gastric juice and food—that the particles of food are constantly changing their relations with each other—and that they are mixed with a quantity of fluid, the gastric juice, liquids that have been taken during the meal, and (as there has generally been observed a large proportion of fluid, even after a dry and solid meal,) I have been led to suspect a synthetic formation of water, from its elements. This mixture is perfectly heterogeneous at first, and is kept in constant agitation, by the *churning* motions of the stomach. If the contents of the stomach be taken out in from thirty minutes to an hour after eating, it will be found to be composed of perfectly formed chyme and particles of food, intimately mixed and blended; sometimes in larger and sometimes in smaller proportions, according to the vigorous or enfeebled state of the digestive organs, or the quantity or quality of aliment taken. Most commonly, if the meal have been moderate, the process of digestion will continue in the portion taken out, when placed on the bath at a proper temperature, and the motions of the stomach imitated.

From the circumstance that the introduction of sponge, tubes, pebbles, &c. by SPALLANZANI and others, excited the discharge of the gastric juice, and from the fact that the gum-elastic tube, in my experiments, produced the same effect, when the stomach was empty and healthy, I infer, that the first effect of aliment on the stomach, is one of *irritation* of the gastric papillæ; thus exciting the discharge of the gastric juice, and stimulating the muscular fibres of the stomach. The vermicular motions, being excited by mechanical irritation, not only carry the ingestæ into all parts of the stomach, and diffuse its mechanical influence throughout the whole inner surface of this organ; but, by this means, they uniformly mix the aliment with the gastric juice, which is constantly being secreted, in proportion to the quantity of food received into the stomach, (unless that be too much for the wants of the economy,) until chymification be completed. Some stimulus seems to be necessary to continue the motions of the stomach, after chymification is accomplished, in order to effect its complete discharge into the lower bowels. And it appears highly probable that the compound fluid of gastric juice and aliment, or chyme, by its acquired acid properties, affords this stimulus, and propagates the contractile motions of this organ, even after the mechanical irritation of the crude food ceases. This fluid acquires new chemical properties, becomes more acid and stimulating, as chymification advances, until it is completed. When it is all transferred to the duodenum, the motions of the stomach cease.

From a number of experiments on rabbits, by Doctor Wilson Philip,† with the view of ascertaining the process of digestion, this gentleman has brought his mind to the conclusion, that when food has been taken at different times, "the new is never mixed with the

† On Digestion.

old food." With every feeling of respect for so valuable and indefatigable a contributor to physiological science, I must beg leave, however, to dissent from this opinion. In many of his experiments, the rabbits were killed soon after the introduction of a fresh quantity of food, and, generally, of a very different kind. The result was, that it was found separate from the old food, which was in an advanced stage of digestion. It was in the centre of the old food, and surrounded by it. This is precisely where a new bolus would be received, and retain its shape and consistence, in some measure, until disturbed, and broken up, by the motions of the stomach. By allowing sufficient time for the action of this organ, it is probable that the line of separation would not have been perceived. Indeed the Doctor concedes that when the second quantity of food was of the same kind as the first, and the rabbit had been left to live for some time, the line of separation was very indistinct. It appears that he fed rabbits on *oats*, and after making them fast for *sixteen* or *seventeen* hours, he fed them as much *cabbage* as they chose to eat, " and killed them at different periods, from *one* to *eight* hours after they had eaten it;" when the line of separation between the new food and that which had been eaten from *eighteen* to *twenty-five* hours before, was, no doubt, *quite distinct.* I confess I know very little about the habits of these animals, as it respects their modes of digestion; but I should be inclined to think that if the " line of separation" between the two portions of food were not sufficiently distinct, it was not for want of *time.* In man, one fifth of the time would have been more than sufficient to have disposed of any reasonable quantity of food.

Comparative physiology, as well as comparative anatomy, is undoubtedly, very useful; but, at the same time, it will not do to make it of general application. The rabbit is a ruminating animal; and is it not probable that the " new food," found in the " small curvature," if it be in fact retained there, is detained for the purpose of regurgitation and re-mastication, before it is digested? If the circumstance be true, and there be no deception in the case, I think this must be the design of the contrivance.

Arguments from analogy may be very plausible, and are certainly very allowable, when the subject presents no other mode; but they are not conclusive. We cannot judge of the mode of digestion in the human stomach by that of animals, particularly the granivorous and ruminating animals. Carnivorous animals most resemble man in their digestive apparatus. One thing is certain, and it is capable of demonstration in the stomach of the subject of these experiments, that old and new food, if they are in the same state of comminution, are readily and speedily mixed in the stomach.

On the subject of exercise or repose, during the digestion of a meal, there has been some diversity of opinion. It has generally been conceded, however, that a state of repose is most favourable to chymification. It has been said that during the digestion of aliment, the *energies* of the system were centred on the stomach, and should not be withdrawn to any distant part; that the stomach becomes a " centre of fluxion," &c. &c. I protest, again, against the use of terms which have no definite meaning. I believe the benefits of science will be better subserved by adhering to facts, and the deductions of experiment, than by the propagation of hypotheses founded on uncertain data. From numerous trials, I am persuaded that moderate exercise conduces considerably to healthy and rapid digestion. The discovery was the result of accident, and contrary to preconceived opinions. I account for it in the following way. Gentle exercise increases the circulation of the system, and the *temperature* of the stomach. This increase of temperature is generally about one and a half degrees. Now, if the gastric juice be a solvent, its action is similar to other chemical solvents, and its rapidity is increased in proportion to the elevation of temperature. Of the reason, I leave others to judge. The effect is certain. Severe and fatiguing exercise, on the contrary, retards digestion. Two reasons present themselves for this—the debility which follows hard labour, of which the stomach partakes; and the depressed temperature of the system, consequent upon perspiration, and evaporation from the surface.

Exercise, sufficient to produce moderate perspiration, increases the secretions from the gastric cavity, and produces an accumulation of a limpid fluid, within the stomach, slightly acid, and possessing the solvent properties of the gastric juice in an inferior degree. This is probably a mixed fluid, a small proportion of which is gastric juice.

Bile is not essential to chymification. It is seldom found in the stomach, except under peculiar circumstances. I have observed that when the use of fat or oily food has been persevered in for some time, there is generally the presence of bile in the gastric fluids. Whether this be a pathological phenomenon, induced by the peculiarly indigestible nature of oily food; or whether it be a provision of nature, to assist the chymification of this particular kind of diet, I have not as yet satisfied myself. Oil is affected by the gastric juice with considerable difficulty. The alkaline properties of the bile may render it more susceptible of solution in this fluid, by altering its chemical character. Irritation of the pyloric extremity of the stomach with the end of the elastic tube, or the bulb of the thermometer, generally occasions a flow of bile into this organ. External agitation, by kneading with the hand, on the right side, over the regions of the liver and pylorus, produces the same effect. It may be laid down as a general rule, however, subject to the exceptions above mentioned, that bile is not necessary to the chymification of food in the stomach. MAGENDIE says, "I believe that, in certain morbid conditions, the bile is not introduced into this organ," (the stomach;) inferring, that in a healthy state, it is always to be found there. There can hardly be a greater mistake. With the exceptions that I have mentioned, it is never found in the gastric cavity, in a state of health; and it is only in "certain morbid conditions" that it is found there.

When bile is found with the gastric juice, the acid taste is diminished, and the flavour of the bile prevails, in proportion to the quantity in the mixture.

The resulting compound of digestion in the stomach, or *chyme*, has been described as "a homogeneous, pultaceous, greyish substance, of a sweetish, insipid taste, slightly acid," &c. In its *homogeneous* appearance, it is invariable; but not in its *colour;* that partakes very slightly of the colour of the food eaten. It is always of a lightish or greyish colour; varying in its shades and appearance, from that of cream, to a greyish, or dark coloured gruel. It is, also, more consistent at one time than at another; modified, in this respect, by the kind of diet used. This circumstance, however, does not affect its homogeneous character. A rich and consistent quantity is all alike, and of the same quality. A poorer and thinner portion is equally uniform in its appearance. Chyme from butter, fat meats, oil, &c. resembles rich cream. That from farinaceous and vegetable diet, has more the appearance of gruel. It is invariably distinctly acid.

The passage of chyme from the stomach is gradual. Portions of chyme, as they become formed, pass out, and are succeeded by other portions. In the early stages, the passage of the chyme into the duodenum, is more slowly effected than in the later stages. At first, it is more mixed with the undigested portions of aliment, and is probably separated with considerable difficulty, by the powers of the stomach. In the later stages, as the whole mass becomes more

more chymified, and fitted for the translation, the process is more rapid; and is accelerated by a peculiar contraction of the stomach, a description of which will be found in the next section. It appears to be a provision of nature, that the chyme, towards the latter stages of its formation, should become more stimulating, and operate on the pyloric extremity of the stomach, so as to produce this peculiar contraction.

After the expulsion of the last particles of chyme, the stomach becomes quiescent, and no more juice is secreted, until a fresh supply of food is presented for its action, or some other mechanical irritation is applied to its internal coat.

Water and alcohol are not affected by the gastric juice. Fluids, of all kinds, are subject to the same exemption, unless they hold in solution or suspension some animal or vegetable aliment. Fluids pass from the stomach very soon after they are received, either by absorption, or through the pylorus.

Since the general adoption of the theory of a specific, solvent fluid, others have been proposed.

M. Montegre, who, it is said, had the power of vomiting at pleasure, performed a series of experiments on the fluids of the stomach, obtained in this way, which induced him to come to a different conclusion on the subject of digestion. " He conceives that what has been supposed to be the gastric juice, is, in fact, nothing but *saliva;* that it possesses no peculiar powers of acting on alimentary matter; that the principal use of the gastric juice is to dilute the food; and that the only action of the stomach consists in 'une absorption vitale et elective,' in which the absorbent vessels, in consequence of their peculiar sensibility, take up certain parts of the food, and reject others."* A complete refutation of the conclusions drawn from the experiments of Montegre, will be found in the fact, which has been tested by more than two hundred examinations and experiments, made by me, on the gastric cavity, that there never exists free gastric juice in the stomach, unless excited by aliment, or other stimulants. The fluid obtained by MONTEGRE was, in all probability, a mixture of saliva (which had been unconsciously swallowed) and the mucus of the stomach. Neither of these secretions are capable of digesting aliment; nor could the peculiar products, generally obtained from the chemical analysis of the gastric juice, be found in them.

The hypotheses proposed by Professors SMITH and JACKSON, of this country, are modifications of MONTEGRE's theory.

The former of these gentlemen supposes that digestion is performed "by the *veins* of the stomach, and by the *liver*." He contends, "that the first step in the process of digestion is effected by capillary veins originating in the villi of the stomach, with absorbing extremities, and terminating in the great branches of the vena portæ;"‡ that this action is continued through the small intestines; that the absorbing

* Note in Bostock's Physiology, vol. 2, p. 384

‡ Essay on Digestion, p. 63.

veins take up the nutrient principles of the food, and reject, as excrementitious, the innutritious part; that these nutrient principles are mixed with the returning blood within the cavity of the abdomen, and are carried into the liver, where the final processes of animalization and conversion into blood are completed.

Professor JACKSON, in a recent work, has proposed a new theory, or rather revived, in some measure, the theory of maceration. His hypothesis, as nearly as can be collected from his work, is as follows:—He supposes that digestion is performed by submitting food to the action of different fluids, each of which has "solvent powers for different principles;"* that the nutrient principles exist already formed in food, and are released from principles that are not required for nutrition, by a species of solution, or maceration. The different fluids, as saliva, mucus from the mouth, throat, stomach, intestines, the bile and pancreatic juice, are the solvents of the different innutritive principles, and separate them from nutriment. He attributes great importance to the action of the saliva; thinks it exercises a "very energetic operation on the food," &c. and denies, altogether, the existence of a specific solvent fluid.

It is unfortunate for the interests of physiological science, that it generally falls to the lot of men of vivid imaginations, and great powers of mind, to become restive under the restraints of a tedious and *routine* mode of thinking, and to strike out into bold and ori-

* Principles of Medicine, founded on the Structure and Functions of the Animal Organism, p. 354.

ginal hypotheses to elucidate the operations of nature, or to account for the phenomena that are constantly submitted to their inspection. The process of developing truth, by patient and persevering investigation, experiment and research, is incompatible with their notions of unrestrained genius. The drudgery of science, they leave to humbler, and more unpretending contributors. The flight of genius is, however, frequently erratic. The bold and original opinions of Brown, for a long time unsettled the practice of medicine; and the later opinions of Montegre and others, have had a like effect on the sister science of physiology. It is, however, a right, which men of genius possess, in common with others, to propose hypotheses, and to support them with such arguments and deductions as they may have in their power to bring. Great caution and circumspection ought, however, to be observed. It is dangerous to unsettle long established truths; for it is difficult to limit the extent of error. The gratification of a *morbid* desire to be distinguished as the propagator of new principles in philosophy, or as the head of a new sect, is not the only result to be expected from such heresies. New opinions or doctrines, whether true or false, will have admirers and followers, and will lead to practical results. And the errors of one man may lead thousands into the same vortex.

These, of course, are designed as general remarks; and I have no wish to apply them, so far as bad motives are inferred, to the highly respectable gentlemen mentioned above. Honest objections, no doubt, are entertained against the doctrine of digestion by the gastric juice. That they are so entertained by these gentlemen, I have no doubt. And I cheerfully concede to them the merit of great ingenuity, talents and learning, in raising objections to the commonly received hypothesis, and ability in maintaining their peculiar opinions. But we ought not to allow ourselves to be seduced by the ingenuity of argument or the blandishments of style. Truth, like beauty, when "unadorned, is adorned the most;" and in prosecuting these experiments and inquiries, I believe I have been guided by its light. Facts are more persuasive than arguments, however ingeniously made, and by their eloquence, I hope I have been able to plead for the support and maintenance of those doctrines, which have had for their advocates such men as SYDENHAM, HUNTER, SPALLANZANI, RICHERAND ABERNETHY, BROUSSAIS, PHILIP, PARIS BOSTOCK, the Heidleburgh and Paris Professors, DUNGLISON, and a host of other luminaries in the science of physiology.

SECTION VI.

OF THE APPEARANCE OF THE VILLOUS COAT, AND OF THE MOTIONS OF THE STOMACH.

THE inner coat of the stomach, in its natural and healthy state, is of a light, or pale pink colour, varying in its hues, according to its full or empty state. It is of a soft, or velvet-like appearance, and is constantly covered with a very thin, transparent, viscid mucus, lining the whole interior of the organ.

Immediately beneath the mucous coat, and apparently incorporated with the villous membrane, appear small, spheroidal, or oval shaped, glandular bodies, from which the mucous fluid appears to be secreted.

By applying aliment, or other irritants, to the internal coat of the stomach, and observing the effect through a magnifying glass, innumerable minute lucid points, and very fine nervous or vascular papillæ, can be seen arising from the villous membrane, and protruding through the mucous coat, from which distills a pure, limpid, colourless, slightly viscid fluid. This *fluid*, thus excited, is invariably distinctly acid. The *mucus* of the stomach is less fluid, more viscid or albuminous, semi-opaque, sometimes a little saltish, and does not possess the slightest character of acidity. On applying the tongue to the mucous coat of the stomach, in its empty, unirritated state, no acid taste can be perceived. When food, or other irritants, have been applied to the villous membrane, and the gastric papillæ excited, the acid taste is immediately perceptible. These papillæ, I am convinced, from observation, form a part of what is called by authors, the villi of the stomach. Other vessels, perhaps absorbing as well as secretory, compose the remainder. That some portion of the villi form the excretory ducts of the vessels, or glands, I have not the least doubt, from innumerable, ocular examinations of the process of secretion of gastric juice. The invariable effect of applying aliment to the internal, but exposed part of the gastric membrane, when in a healthy condition, has been the exudation of the solvent fluid, from the above mentioned papillæ.—Though the *apertures* of these vessels could not be seen, even with the assistance of the best microscopes that could be obtained; yet the points from which the fluid issued was clearly indicated by the gradual appearance of innumerable, very fine, lucid specks, rising through the transparent mucous coat, and seeming to burst, and discharge themselves upon the very points of the papillæ, diffusing a limpid, thin fluid over the whole interior gastric surface. This appearance is conspicuous only during alimentation, or chymification. These lucid points, I have no doubt, are the termination of the excretory ducts of the gastric vessels or glands, though the closest and most accurate observation may never be able to discern their distinct apertures.

The fluid, so discharged, is absorbed by the aliment in contact, or collects in small drops, and trickles down the sides of the stomach, to the more depending parts, and there mingles with the food, or whatever else may be contained in the gastric cavity. This fluid, the efficient cause of digestion—the true gastric juice of SPALLANZANI, I have no doubt—has generally been obtained, for experiment, by mechanical irritation of the internal coat of the stomach, produced by the introduction of a gum-elastic tube, through which it has been procured.

The gastric juice never appears to be accumulated in the cavity of the stomach while fasting; and is seldom, if ever, discharged from its proper secerning vessels, except when excited by the natural stimulus of aliment, mechanical irritation of tubes, or other excitants. When aliment is received, the juice is given out in exact proportion to its requirements for solution, except when more food has been taken than is necessary for the wants of the system.

When mechanical irritation by a non-digestible substance, as the elastic tube, stem of the thermometer, &c. has been used, the secretion is probably less than when the irritation has been produced by such substances, as are readily dissolved in the gastric juice. Alimentary stimulus, when taken into the stomach, is diffused over the whole villous surface, and excites the gastric vessels, generally, to excrete their fluids copiously; whereas the irritation of tubes, &c.

is local, and produces only a partial excitement of the vessels, and a scanty flow of the gastric juice. Hence, the slowness in obtaining the clear fluid from the empty stomach, through the tube. I have never, on numerous trials, been able to obtain, at any one time, more than one and a half, or two ounces of this fluid, after the stomach had disposed of its alimentary matters, however long the period of abstinence had been. The discharge of this small quantity has generally been excited by the introduction of the tube. Ten, fifteen, or more minutes, were necessary to collect even this small quantity. Whenever fluid was obtained in larger quantity, as was sometimes the case, it invariably contained more than the usual quantity of mucus.

On viewing the interior of the stomach, the peculiar formation of the inner coats are distinctly exhibited. When empty, the rugæ appear irregularly folded upon each other, almost in a quiescent state, of a pale pink colour, with the surface merely lubricated with mucus. On the application of aliment, the action of the vessels is increased; the colour brightened; and the vermicular motions excited. The small gastric papillæ begin to discharge a clear, transparent fluid, (the alimentary solvent,) which continues abundantly to accumulate, as aliment is received for digestion.

If the mucous covering of the villous coat be wiped off, with a sponge or handkerchief, during the period of chymification, the membrane appears roughish, of a deep pink colour at first; but in a few seconds, the follicles and fine papillæ begin to pour out their respective fluids, which, being diffused over the parts abraded of mucus, restore to them their peculiar soft and velvet-like coat, and pale pink colour, corresponding with the undisturbed portions of the membrane; and the gastric juice goes on accumulating, and trickles down the sides of the stomach again.

If the membrane be wiped off when the stomach is empty, or during the period of fasting, a similar roughness, and deepened colour appear, though in a less degree; and the mucous exudation is more slowly restored. The follicles appear to swell more gradually. The fluids do not accumulate in quantity sufficient to trickle down, as during the time of chymification. The mucous coat only, appears to be restored.

The foregoing, I believe to be the natural appearances of the internal coat of the stomach, in a healthy condition of the system.

In disease, or partial derangement of the healthy function, this membrane presents various, and essentially different appearances.

In febrile diathesis, or predisposition, from whatever cause—obstructed perspiration, undue excitement by stimulating liquors, overloading the stomach with food—fear, anger, or whatever depresses or disturbs the nervous system—the villous coat becomes sometimes red and dry, at other times, pale and moist, and loses its smooth and healthy appearance; the secretions become vitiated, greatly diminished, or entirely suppressed; the mucous coat scarcely perceptible; the follicles flat and flaccid, with secretions insufficient to protect the vascular and nervous papillæ from irritation.

There are sometimes found, on the internal coat of the stomach, eruptions, or deep red pimples; not numerous, but distributed, here and there, upon the villous membrane, rising above the surface of the mucous coat. These are at first sharp pointed and red; but frequently become filled with white purulent matter. At other times, irregular, circumscribed, red patches, varying in size or extent, from half an inch to an inch and a half in circumference, are found on the internal coat. These appear to be the effect of congestion in the minute blood vessels of the stomach. There are, also, seen at times, small aphthous crusts, in connection with these red patches. Abrasions of the lining membrane, like the rolling up of the mucous coat into small shreds or strings, leaving the papillæ bare, for an indefinite space, is not an uncommon appearance.

These diseased appearances, when very slight, do not always affect, essentially, the gastric apparatus. When considerable, and, particularly, when there are corresponding symptoms of disease, as dryness of the mouth, thirst, accelerated pulse, &c. no gastric juice can be extracted, not even on the application of alimentary stimulus. Drinks received, are immediately absorbed, or otherwise disposed of; none remaining in the stomach ten minutes after being swallowed. Food, taken in this condition of the stomach, remains undigested for twenty-four or forty-eight hours, or more, increasing the derangement of the whole alimentary canal, and aggravating the general symptoms of disease.

After excessive eating or drinking, chymification is retarded; and, although the appetite be not always impaired at first, the fluids become acrid and sharp, excoriating the edges of the aperture; and almost invariably produce aphthous patches, and the other indications of a diseased state of the internal membrane, mentioned above. Vitiated bile is also found in the stomach under these circumstances; and flocculi of mucus are much more abundant than in health.

Whenever this morbid condition of the stomach occurs, with the usual accompanying symptoms of disease, there is generally a corresponding appearance of the tongue. When a healthy state of the stomach is restored, the tongue invariably becomes clear.

Motions of the Stomach.—With the *anatomy* of this organ, I have, at present, nothing to do. It does not come within the limits which I have prescribed to myself. Its *motions*, as comprising a part of the process of digestion, I have endeavoured to observe as accurately as practicable, and I give the result.

The human stomach is furnished with muscular fasciculi, so arranged as to shorten its diameter in every direction. By the alternate contraction and relaxation of these bands, a great variety of motion is induced on this organ, sometimes transversely, and at other times longitudinally. These alternate contractions and relaxations, when affecting the transverse diameter, produce what are called *vermicular* or *peristaltic* motions. The effect of the contraction of the longitudinal fibres, is to approximate the splenic and pyloric extremities. When they all act together, the effect is to lessen the cavity of the stomach, and to press upon the contained aliment, if there be any in the stomach. These motions not only produce a constant disturbance, or *churning* of the contents of this organ, but they compel them, at the same time, to revolve around the interior, from point to point, and from one extremity to the other. In addition to these motions, there is a constant agitation of the stomach, produced by the respiratory muscles.

These contractions and relaxations of the muscular fasciculi, do not observe any very *exact* mode. Their motions are modified by various circumstances, such as the stimulant or non-stimulant property of the ingestæ, the healthy or unhealthy state of the internal coat of the stomach; by exercise, and by repose, &c. &c.

The ordinary course and direction of the revolutions of the food, are first, after passing the œsophageal ring, from right to left, along the small arch; thence, through the large curvature, from left to right. The bolus, as it enters the cardia, turns to the left; passes the aperture; descends into the splenic extremity; and follows the great curvature towards the pyloric end. It then returns, in the course of the smaller curvature, makes its appearance again at the aperture, in its descent into the great curvature, to perform similar revolutions.

Such I have ascertained to be the revolutions of the contents of the stomach, from being able to identify particular portions of food, and from the fact, that the bulb of the thermometer, which has been frequently introduced during chymification, invariably indicates the same movements. These revolutions are completed in from one to three minutes. They are probably induced, in a great measure, by the circular or transverse muscles of the stomach, as indicated by the spiral motion of the stem of the thermometer, both in descending to the pyloric portion, and ascending to the splenic.† These motions are slower at first than after chymification has considerably advanced.

While these revolutions of the contents of the stomach are progressing, the trituration or agitation is also going on. There is a perfect admixture of the whole ingestæ, during the period of alimentation and chymification. There is nothing of the distinct lines of separation between old and new food, and peculiar central or peripheral situation of crude, as distinguished from chymified aliment, said to have been observed by PHILIP, MAGENDIE and others, in their experiments on dogs and rabbits, to be seen in the human stomach; at least in that of the subject of these experiments. The whole contents of the stomach, until chymification be nearly complete, exhibit a heterogeneous mass of solids and fluids; hard

† The terms "descending" and "ascending," are used here, as well as in many other places, relatively; because the examinations were generally made while the man was lying on his right side.

and soft; coarse and fine; crude and chymified; all intimately mixed, and circulating promiscuously through the gastric cavity, like the mixed contents of a closed vessel, gently agitated, or turned in the hand.

If a mouthful of some tenacious food be swallowed, after digestion is considerably advanced, it will be seen passing the opening, to the great curvature; and in the course of about one and a half or two minutes, it will reappear, with the general circulating contents, more or less broken to pieces, or divided into smaller pieces; and very soon loses its identity. This agitating motion has the effect, and is undoubtedly designed, to break up the bolus, as well as to separate the external and chymified portion of the particles of food, and allow the undigested portions to come in contact with the gastric juice, their proper solvent. If the motions were simply revolutionary, the central portions would retain their situation, until the outer, or chymified part, had passed into the duodenum, in successive parcels; which, it is evident, would very much retard the process of digestion.

As the food becomes more and more changed from its crude to its chymified state, the acidity of the gastric fluids is considerably increased; more so in vegetable than animal diet; and the general contractile force of the muscles of the stomach is augmented in every direction; giving the contained fluids an impulse towards the pylorus.

It is probable, that from the very commencement of chymification—from the time that food is received stomach—until that organ becomes empty, portions of chyme are constantly passing into the duodenum, through the pyloric orifice, as the mass is presented at each successive revolution. I infer this, from the fact that the volume is constantly decreasing. This decrease of volume, however, is slow at first; but is rapidly accelerated towards the conclusion of digestion, when the whole mass becomes more or less chymified. This accelerated expulsion appears to be effected by a peculiar action of the transverse muscles, or rather of the *transverse band*, as described by SPALLANZANI, HALLER, COOPER, Sir E. HOME, and others, in their experiments on animals. This band is situated near the commencement of the more conically shaped part of the pyloric extremity, three or four inches from the smaller end. In attempting to pass a long glass thermometer tube, through the aperture, into the pyloric portion of the stomach, during the latter stages of digestion, a forcible contraction is first perceived at this point, and the bulb is stopped. In a short time, there is a gentle relaxation, when the bulb passes without difficulty, and appears to be drawn, quite forcibly, for three or four inches, towards the pyloric end, It is then released, and forced back, or suffered to rise again; at the same time, giving to the tube a circular, or rather spiral motion, and frequently revolving it completely over. These motions are distinctly indicated, and strongly felt, in holding the end of the tube between the thumb and finger; and it requires a pretty forcible grasp to prevent it from slipping from the hand, and being drawn suddenly down to the pyloric extremity.

When the tube is left to its own direction, at these periods of contraction, it is drawn in, nearly its whole length, to the depth of ten inches: and when drawn back, requires considerable force, and gives to the fingers the sensation of a strong *suction* power, like drawing the piston from an exhausted tube. This ceases as soon as the relaxation occurs, and the tube rises again, of its own accord, three or four inches, when the bulb seems to be obstructed from rising further; but if pulled up an inch or two, through the stricture, it moves freely in all directions in the cardiac portions, and mostly inclines to the splenic extremity, though not disposed to make its exit at the aperture.

Above the contracting band, and towards the splenic portion of the stomach, the suction or grasping motion is not perceptible; but when the bulb is pushed down to this point, it is distinctly felt to be grasped, and confined in its movements.

These peculiar motions and contractions continue until the stomach is perfectly empty, and not a particle of food or chyme remains; when all becomes quiescent again.

If the bulb of the thermometer be suffered to be drawn down to the pyloric extremity, and detained there for a short time, or if the experiment be repeated too frequently, it causes severe distress, and a sensation like cramp, or spasm, which ceases on withdrawing the tube; but leaves a sense of soreness and tenderness at the pit of the stomach.

These peculiar contractions and relaxations, mentioned above, succeed each other, at irregular intervals, of from two to four or five minutes. Simultaneously with the contractions, there is a general shortening of the fibres of the stomach. This organ contracts upon itself in every direction; and its contents are compressed with much force. The valvular portion of the stomach is firmly thrust into the aperture; closing the orifice; preventing the egress of aliment; and obstructing the view of the interior. During the intervals of relaxation, the rugæ perform their vermicular actions, the undulatory motions of the fluids continue, and the alimentary and chymous mass appear, revolving as before, promiscuously mixed, through the splenic and cardiac portions.

All these facts, taken together, will, I think, rationally admit of the following explanation. The longitudinal muscles of the whole stomach, with the assistance of the transverse ones of the splenic and central portions, carry the contents into the pyloric extremity. The circular or transverse muscles contract progressively, from left to right. When the impulse arrives at the *transverse band*, this is excited to a more forcible contraction, and, closing upon the alimentary matter and fluids, contained in the pyloric end, prevents their regurgitation. The muscles of the pyloric end, now contracting upon the contents detained there, separate and expel some portion of the chyme. It appears that the crude food excites the contractile power of the pylorus, so as to prevent its passage into the duodenum, while the thinner, chymified portion is pressed through the valve, into the intestine. After the contractile impulse is carried to the pyloric extremity, the circular band, and all the transverse muscles, become relaxed, and a contraction commences in a reversed direction, from right to left, and carries the contents again to the splenic extremity, to undergo similar revolutions.

It would appear, then, that the discharge of the chyme from the stomach, is effected by *mechanical* impulse. But, I confess, I do not like to give an opinion. I state the circumstances as they have occurred. The idea of mechanical force, I admit, is liable to objection; but, perhaps, not more so than that of the *selecting* power of the pylorus. Whatever bias I may have in favour of the former method, has been forced upon me by the deductions of experiment and observation.

SECTION VII.

Of Chymification, and Uses of the Bile and Pancreatic Juice.

AS food becomes chymified by the gastric juice, the contractile motions of the stomach send it into the duodenum, to receive further changes, preparatory to its assimilation to the circulating fluids of the system, by the lacteal absorbents. It is at first slowly received into this organ from the stomach; but during the later stages of chymification, its transmission becomes more accelerated. The duodenum is so constituted, that the passage of the chyme through it, is considerably retarded; and, hence, in some pathological conditions of the system, the pressure on that organ from repletion, is considerable; and frequently produces great pain and distress.

The vermicular motions of this and the other intestines, are propagated from the stomach, and are continued, after this organ has discharged all its contents, by the contained fluids, until the whole becomes assimilated. They are more or less rapid, varying at different sections of the canal; of which it is not necessary to particularize. These motions are excited by the stimulus of the chyme, and occur at intervals, on the introduction of each quantity passed through the pylorus.

The chymous mass is not changed until it arrives at, or passes the mouth of the ductus cholodochus, when the liver and pancreas are excited to discharge their respective fluids. These mix with the chyme, and produce an essential alteration in its sensible and chemical properties. At this point, the lacteal absorbents commence.

That the change from a *chymous* to a *chylous* stage is effected by the operation of the bile and pancreatic juice, there can be no doubt. Of the nature of this change, there is some diversity of opinion.— Chyle is generally described as "a white, opaque substance, considerably resembling cream in its aspect and physical properties;"* though it is *said* to vary slightly, according to the kind of aliment which had been used. It is my impression, however, that pure chyle, taken from the lacteals of a healthy subject, and produced by natural food, is invariably the same substance in the same individual. Changes that have been observed, must be reckoned as the effect of a pathological state of the system, or the absorption of a non-digesting substance. Medicines and other substances, which are not capable of digestion, are sometimes taken up by the lacteal absorbents, and may produce an alteration in the physical and chemical properties of chyle. It is possible

* Bostock's Physiology, vol. 2d, p. 392.

that a small proportion of oil may escape the action of the digestive apparatus, be absorbed by the lacteals, and produce the opaque, white colour, mentioned by authors, as sometimes appearing. Countenance is given to this suggestion, by the fact, that the more opaque coloured parts of chyle are found floating on the surface; and that it is always discovered after the ingestion of oily food. At other times, it is uniform in its colour and consistence, whatever colouring matter may have been contained in the food.

I wish to be understood to say, that every species of aliment produces the same *kind* of nutrient principles. With the view of attempting an investigation of this subject, as has been previously mentioned, I instituted some imperfect experiments and examinations. For the result, see Experiments, Second Series, from 47th to 56th. By the addition of bile and dilute muriatic acid, and subsequently pancreatic juice, to chyme formed in the artificial way, as well as in the stomach, it separated into three distinct parts, a reddish brown sediment at the bottom, a whey coloured fluid in the centre, and a creamy pellicle at the top. Each repetition of the experiment produced a *similar* result; though not *exactly* alike in all. The central portion, I suspect to be imperfectly formed chyle. The sediment, from its appearance, and the coarseness of its particles, I judge, is incapable of being acted on, or taken up, by the absorbents: the creamy or oily pellicle is not only liable to the same objection, but is in too small proportion to the ingesta. The fluid part is fitted, by its fluidity, for the ready action of the absorbents; and is, moreover, in sufficient quantity for the purposes of nutrition. The change of colour and consistence is, probably, effected in the lacteal glands and vessels. The sediment and pellicle, I apprehend, are both excrementitial. The "irregular filaments," attached to the valvulæ conniventes, mentioned by MAGENDIE, and which he concluded to be imperfectly formed chyle, were, undoubtedly, portions of the creamy pellicle, found in the experiments referred to.

But what is the nature of the changes effected in the duodenum? Aliment, after being introduced into the stomach, is dissolved in the gastric juice, and forms a new compound with this fluid. The constituent elements of food are various. When compounded with the gastric juice, they may, nevertheless, be said to form a *simple* compound, or a *gastrite* of *aliment*. I am indifferent about terms; and this will as well convey my meaning as any other. When this *gastrite* is introduced into the duodenum, and mixed with the hepatic and pancreatic fluids, are we not warranted, from all the facts that have been observed, in saying, that there is a general play of chemical affinities in that organ, separating the nutrient principles, and forming various new compounds from the elements of each? The chymous mass changes its colour, and loses its acidity. There is a sensible extrication of gas, as observed by MAGENDIE, and others.* In the stomach, oxigen is found mixed with a small proportion of hydrogen. In the intes-

* The escape of gas is generally observable in mixing these fluids with chyme, in my experiments.

tines, an increased proportion of hydrogen exists, with carbonic acid, nitrogen, &c.; but no oxygen. Does not the acid of the chyme unite with the alkalis of the bile, and form new compounds? And do not other equally important changes take place? This subject, I confess, is obscure, and perhaps will not admit of a very perfect investigation.

The constant agitation which is maintained in the intestines, preserves the chyle in a state of perfect admixture with the other fluids, until absorption has taken place. By standing at rest, the separation, mentioned above, is evident and perfect.

It has been supposed that the mucus of the intestines has some agency in the formation of chyle. But I am disposed to think, with Professor DUNGLISON, and others, that the use of the mucus is to lubricate the internal coat of the intestines, and, perhaps, to dilute their contents.

It has been suggested that digestion can be perfected in the duodenum and lower bowels, when the food has not been submitted to the action of the stomach and its fluids. In two experiments by MAGENDIE, one failed, and the other was attended with partial success. Too much reliance ought not to be placed on experiments, that require such severe and cruel vivisections, as were resorted to in these cases. It is possible, as suggested by DUNGLISON, that the presence of crude aliment in the duodenum, may excite the discharge of gastric juice in the stomach, its expulsion into the duodenum, and its consequent action on the food, before it is affected by the bile and pancreatic juice. Or, it may be that the upper part of the duodenum is furnished with vessels, which secrete a fluid similar to gastric juice.

Experiments have also been instituted with the view of ascertaining, whether chyle can be formed without the admixture of the hepatic and pancreatic fluids, with various results. BRODIE ascertained, by tying the ductus communis cholodochus in young cats, that the process of chylification was prevented, and that no chyle was found in the intestines. MAGENDIE, LEURET and LASSAIGNE, on tying this duct, discovered matter of "a rosy yellow colour," which afforded, on analysis, the same constituents of chyle, although the animals, which were the subjects of the operation, *had been kept some time without food.* There is certainly an apparent discordance in these reports. But, it is possible, they may be explained, and reconciled. It is well known that the absorbents are active during a protracted fast, (as in these last experiments) and are constantly taking up the cellular substance, for the purpose of supplying the blood vessels with these broken up solids of the system. Emaciation is the effect of absorption. The lacteals, like other absorbents, have, undoubtedly, their appropriate stimulus; but if that be withholden, they will feed on other substances, the cellular and other solid parts, within their reach. If such be the case, it will account for the rosy coloured fluid, found in the lacteals, by MAGENDIE and others.

EXPERIMENTS

AND

OBSERVATIONS, &C.

EXPERIMENTS, &c.

FIRST SERIES.

Experiment 1.

August 1, 1825. At 12 o'clock, M., I introduced through the perforation, into the stomach, the following articles of diet, suspended by a silk string, and fastened at proper distances, so as to pass in without pain—viz. :—a piece of high seasoned *a la mode beef;* a piece of *raw, salted, fat pork ;* a piece of *raw, salted, lean beef ;* a piece of *boiled, salted beef ;* a piece of *stale bread ;* and a bunch of *raw, sliced cabbage ;* each piece weighing about two drachms ; the lad continuing his usual employment about the house.

At 1 o'clock, P. M., withdrew and examined them —found the *cabbage* and *bread* about half digested: the pieces of *meat* unchanged. Returned them into the stomach.

At 2 o'clock, P. M., withdrew them again—found the *cabbage*, *bread*, *pork*, and *boiled beef*, all cleanly digested,† and gone from the string; the other pieces of

† These Experiments are inserted here, as they were originally taken down in my Note Book, with very little alteration of phraseology, and none of the sense. Subsequent experiments have sometimes convinced me of errors in former ones. When this has been the case, I have generally made the corrections in the way of remarks, or observations, as in this experiment.

meat but very little affected. Returned them into the stomach again.

At 2 o'clock, P. M., examined again—found the *a la mode beef* partly digested: the *raw beef* was slightly macerated on the surface, but its general texture was firm and entire. The smell and taste of the fluids of the stomach were slightly rancid; and the boy complained of some pain and uneasiness at the breast. Returned them again.

The lad complaining of considerable distress and uneasiness at the stomach, general debility and lassitude, with some pain in his head, I withdrew the string, and found the remaining portions of aliment nearly in the same condition as when last examined; the fluid more rancid and sharp. The boy still complaining, I did not return them any more.

August 2. The distress at the stomach and pain in the head continuing, accompanied with costiveness, a depressed pulse, dry skin, coated tongue, and numerous white spots, or pustules, resembling coagulated lymph, spread over the inner surface of the stomach, I thought it advisable to give medicine; and, accordingly, dropped into the stomach, through the aperture, half a dozen *calomel pills*, four or five grains each; which, in about three hours, had a thorough cathartic effect, and removed all the foregoing symptoms, and the diseased appearance of the inner coat of the stomach. The effect of the medicine was the same as when administered in the usual way, by the mouth and œsophagus, except the nausea commonly occasioned by swallowing pills.

This experiment cannot be considered a fair test of the powers of the gastric juice. The cabbage, one of the articles which was, in this instance, most speedily dissolved, was cut into small, fibrous pieces, very thin, and necessarily exposed, on all its surfaces, to the action of the gastric juice.

The stale bread was porous, and, of course, admitted the juice into all its interstices; and probably fell from the string as soon as softened, and before it was completely dissolved. These circumstances will account for the more rapid disappearance of these substances, than of the pieces of meat, which were in entire solid pieces when put in. To account for the disappearance of the fat pork, it is only necessary to remark, that the fat of meat is always resolved into oil, by the warmth of the stomach, before it is digested. I have generally observed that when he has fed on fat meat or butter, the whole superior portion of the contents of the stomach, if examined a short time after eating, will be found covered with an oily pellicle. This fact may account for the disappearance of the pork from the string. I think, upon the whole, and subsequent experiments have confirmed the opinion, that fat meats are less easily digested than lean, when both have received the same advantages of comminution. Generally speaking, the looser the texture, and the more tender the fibre, of animal food, the easier it is of digestion.

This experiment is important, in a pathological point of view. It confirms the opinion, that undigested portions of food in the stomach produce all the phenomena of fever; and is calculated to warn us of the danger of all excesses, where that organ is concerned. It also admonishes us of the necessity of a perfect comminution of the articles of diet.

Experiment 2.

Aug. 7. At 11 o'clock, A. M., after having kept the lad fasting, for seventeen hours, I introduced the glass tube of a Thermometer (Fahrenheit's) through the perforation, into the stomach, nearly the whole length of the stem, to ascertain the natural warmth of the stomach. In fifteen minutes, or less, the mercury rose to 100°, and there remained stationary. This I determined by marking the height of the mercury on the glass, with ink, as it stood in the stomach, and then withdrawing it, and placing it on the gaduated scale again.

I now introduced a gum-elastic (caoutchouc) tube, and drew off one ounce of pure gastric liquor, unmixed with any other matter, except a small proportion of mucus, into a three ounce vial. I then took a solid piece of *boiled, recently salted beef*, weighing three drachms, and put it into the liquor in the vial; corked the vial tight, and placed it in a saucepan, filled with water, raised to the temperature of 100°, and kept at that point, on a nicely regulated sand bath. In *forty minutes* digestion had distinctly commenced over the surface of the meat. In *fifty minutes* the fluid had become quite opaque and cloudy ; the external texture began to separate and become loose. In *sixty minutes*, chyme began to form.

At 1 o'clock, P. M., (digestion having progressed with the same regularity as in the last half hour,) the cellular texture seemed to be entirely destroyed, leaving the muscular fibres loose and unconnected, floating about in fine small shreds, very tender and soft.

At 3 o'clock, the muscular fibres had diminished one half, since last examination, at 1 o'clock.

At 5 o'clock, they were nearly all digested; a few fibres only remaining.

At 7 o'clock, the muscular texture was completely broken down; and only a few of the small fibres floating in the fluid.

At 9 o'clock, every part of the meat was completely digested.

The gastric juice, when taken from the stomach,

was as clear and transparent as water. The mixture in the vial was now about the colour of whey. After standing at rest a few minutes, a fine sediment, of the colour of the meat, subsided to the bottom of the vial.

Experiment 3.

At the same time that I commenced the foregoing experiment, I suspended a piece of *beef*, exactly similar to that in the vial, (Ex. 2d) into the stomach, through the aperture.

At 12 o'clock, M., withdrew it, and found it about as much affected by digestion as that in the vial; there was little or no difference in their appearance. Returned it again.

At 1 o'clock, P. M., I drew out the string; but the meat was all completely digested, and gone.

The effect of the gastric juice on the piece of meat, suspended in the stomach, was exactly similar to that in the vial, only more rapid after the first half hour, and sooner completed. Digestion commenced on, and was confined to, the surface entirely, in both situations. Agitation accelerated the solution in the vial, by removing the coat that was digested on the surface; enveloping the remainder of the meat in the gastric fluid; and giving this fluid access to the undigested portions.

Experiment 4.

Aug. 8. At 9 o'clock, A. M., I drew off an ounce and a half of gastric juice, into a three ounce vial; suspended two pieces of *boiled chicken*, from the breast and back, into it, and placed it in the same situation and temperature as in the second experiment; observing the same regularity and minuteness.

Digestion commenced and progressed much the same, as in the second experiment, but rather slower; the *fowl* appearing to be more difficult of digestion than the *flesh*. The texture of the *chicken* being closer than that of the *beef*, the gastric juice appeared not to insinuate itself into the interstices of the muscular fibre, so readily as into the beef; but operated entirely upon the outer surface, dissolving it as a piece of gum arabic is dissolved in the mouth, until the last particle was digested.

The colour of the fluid, after digesting the chicken, was of a greyish white, and more resembled a milky fluid than whey, which was the colour of the chyme from the beef.

The contents of both vials, kept perfectly tight, remained free from any fœtor, acidity, or offensive smell or taste, from the time of the experiments, (7th and 8th August,) to the 6th of September; at which time, that containing the solution of *boiled beef*, became very offensive and putrid; while that containing the chyme from the *boiled chicken*, was perfectly bland and sweet. Both were kept in exactly similar situations.

It is, perhaps, unnecessary to make any comments on the result of the above experiments. Each one will make up his opinion from the facts. *These* demonstrate, at least, that the stomach secretes a fluid which possesses *solvent* properties. The change in the solid substances is effected too rapidly to be accounted for on the principle of either maceration or putrefaction. I shall be able to show, in some of the following experiments, that aliment undergoes the same changes in the stomach, as is effected in the mode here adopted.

The young man, who was the subject of these experiments, left me about this time, (September, 1825) and went to Canada, the place of his former residence. The experiments were consequently suspended.

EXPERIMENTS, &C.

SECOND SERIES.

FORT CRAWFORD, *Upper Mississippi*,
June 20th, 1829.

ALEXIS ST. MARTIN having returned from Canada, after an absence of nearly four years, with his stomach in the same, or very similar condition, as when he left me in September, 1825, I continued to prosecute the gastric experiments, which were commenced before he left me.

With a view to ascertain the variations of temperature, if any there were, in the interior of the sto mach, under different circumstances and conditions of the system, and vicissitudes of the atmosphere, I instituted the following experiments.

Experiment 1.

Dec. 6, 1829. At 9 o'clock, A. M., I introduced the glass tube of a Thermometer (Fahrenheit's) through the artificial opening into the stomach, in a healthy and empty condition, nearly the whole length of the stem. In six or eight minutes, the mercury became stationary, at 94°. Weather cloudy, damp, and almost raining—ground wet, muddy and thawing. Wind S. and mild. Thermometer, in a North oxposure, 63°. Commenced raining at 11 o'-clock, A. M., and continued all day, with oppressive atmosphere.

Experiment 2.

Dec. 7. Introduced Thermometer at the same hour as yesterday—circumstances of stomach the same. Mercury at 96°. Weather cloudy—Atmosphere damp—Wind N. W. and light—Th: 27°.

Experiment 3.

Dec. 8. Introduced Thermometer at 9 o'clock, A. M.—circumstances of stomach same as yesterday. Mercury stationary at 99°. Weather clear—Atmosphere dry—Wind S. W. and light—Th: 13°.

Experiment 4.

Dec. 9. Introduced Thermometer at 9 o'clock, A. M.—circumstances similar. Mercury stationary at 99°. Weather clear—Atmosphere dry—Wind W. and light—Th: 10°.

Experiment 5.

Jan. 24, 1830. Introduced Thermometer at 3 o'-clock, P. M. Weather clear and cold—Th: 8° below o.—Wind N. W. and light—stomach empty, and coats healthy. Mercury stationary at 100°

Experiment 6.

Jan. 25. Introduced Thermometer at 8 o'clock, A. M. Weather clear—Wind S. W. and light—Th: 2°—Stomach empty. Mercury stationary at 100°.

At 10 o'clock, A. M., (one hour after eating a breakfast of pork and bread,) introduced Thermometer again. Mercury stationary at 100°, as at 9 o'-clock, before eating.

Experiment 7.

March 17. At 10 o'clock, A. M., introduced Thermometer. Weather rainy and foggy—Wind S. W. and light—Th: 38°—Stomach empty, having eaten nothing since 7 o'clock last evening. Mercury stationary at 99°.

Experiment 8.

March 18. At 8 o'clock, A. M., introduced Thermometer. Mercury stationary at 100°. Weather clear—Wind N. W.—Th : 6°.

At 9 o'clock, breakfasted on meat, biscuit and butter, with coffee. Temperature of the stomach, immediately before eating, 100° : thirty minutes after finishing breakfast the temperature had risen to 102°. Digestion rapidly advancing.

It appears, from the above experiments, that the variations of the atmosphere produce effects upon the temperature of the stomach ; a dry atmosphere increasing, and a humid one diminishing it. What would be the effect of copious perspiration, in warm weather, on the temperature of the stomach ? Would that of the interior of this organ be lessened by evaporation ? I regret that sufficient experiments have not been made, fully to satisfy these inquiries. From one or two experiments, it would seem, that the heat of the stomach was increased during the active period of digestion. This, however, was probably owing to exercise, immediately after eating, though not particularly observed and noted at the time. Subsequent experiments have not shown this result. On the contrary, the temperature has been found to be the same, in its full and empty state.

The ordinary temperature of the healthy stomach, may be fairly estimated at 100°, Fahrenheit. Some allowance ought, probably, to be made, in these experiments, for imperfect instruments. It appears, from subsequent examinations, that there is probably some difference of temperature in different regions of the stomach, it being higher at the pyloric than at the splenic end. See subsequent experiments and observations.

To ascertain whether the Gastric Juice be *accumulated* in the stomach, during periods of fasting, or even from the immediate and direct influence of hunger, I made the following experiments.

Experiment 9.

Dec. 5, 1829. At 8 o'clock, A. M., after twelve hours abstinence from either food or drinks, I introduced, at the perforation, a gum-elastic tube, and drew off a drachm or two only of the gastric juice.—There was no accumulation in the stomach.

Experiment 10.

Dec. 12. At 3 o'clock, P. M., introduced tube—could procure two or three drachms only—this was secreted on the irritation of the tube. Stomach contained none in a free state.

Experiment 11.

Dec. 14. At 10 o'clock, P. M., after eighteen hours fasting, introduced tube, and drew off one and a half ounces of gastric juice. It was clear, and almost transparent; tasted a little saltish and acid, when applied to the tongue, similar to thin mucilage of gum arabic, slightly acidulated with muriatic acid. There was no accumulation in the stomach when the tube was introduced.

Experiment 12

March 13, 1830. At 10 o'clock, A. M.—stomach empty—introduced tube; but was unable to obtain any gastric juice. On the application of a few crumbs of bread to the inner surface of the stomach, the juice began slowly to accumulate, and flow through the tube. The crumbs of bread adhered to the mucous coat, soon became soft, and began to dissolve and digest, On viewing the villous membrane *before* applying the bread crumbs, the mucous coat and subjacent follicles only, could be observed ; but immediately *afterwards*, small, sharp papillæ, and minute lucid points, situated in the interstices of, and less than, the mucous follicles, became visible; from which exuded a clear, transparent liquor. It then began to run through the tube.

Experiment 13.

March 18. At 6 o'clock, P. M., after fasting from 8 o'clock, A. M., introduced tube—obtained one and a half ounces gastric juice, after having kept up the irritation, by moving the tube from point to point, for twelve or fifteen minutes. No accumulation of free juice in the stomach.

Experiment 14.

Jan. 26, 1831. At 9 o'clock, A. M.—stomach empty—extracted one ounce gastric juice, slowly through the tube, with the usual admixture of mucus. Introduced food, and it began directly to flow more freely through the tube.

Experiment 15.

Jan. 27. At 8 o'clock, A. M.—stomach empty—introduced elastic tube, and obtained one and a half drachms of gastric juice, by very slow distillation. Applied crumbs of bread to the villous coat, and the juice began immediately to flow freely through the tube.

Experiment 16.

March 6. At 8 o'clock, A. M., extracted two ounces gastric juice, and added it to two ounces of *Madeira wine.* No visible change was produced—no coagulæ formed. They united, like pure water and wine. Heat produced no other effect.

Experiment 17.

March 7. At 6 o'clock P. M.—stomach empty—extracted one and a half ounces of juice, and mixed it with the same quantity of *Jamaica spirits.* Effect same as with wine.

Experiment 18.

March 8. At 8 o'clock, A. M.—stomach empty—extracted one and a half ounces of gastric juice.

Experiment 19.

March 12. At 9 o'clock, A. M.—stomach empty—extracted one and a half ounces of gastric juice. Put this in a bottle.

Experiment 20.

March 13. At 11 o'clock, A. M.—stomach empty—extracted two ounces of juice.

Experiment 21.

March 14. At 12 o'clock, M.—stomach empty—extracted two ounces of juice.

Experiment 22.

March 15. At 4 o'clock P. M.—stomach empty—extracted one and a half ounces gastric juice.

Experiment 23.

March 16. At 5 o'clock P. M., introduced tube—could obtain no clear gastric juice. A little acrid fluid and frothy mucus, only, could be extracted. Villous membrane red and dry. St. Martin complained of some head ache, pain and distress about the scrobiculus cordis, lassitude and loss of appetite. Directed him to take half an ounce of *tincture of aloes and myrrh*, at 9 o'clock, P. M. This moved his bowels several times next morning. Little or no change was apparent in the appearance of the inner coat of the stomach; if any, it was a little more moist, and a shade paler, after the operation of the tincture. Gastric juice could again be obtained, but in less than usual quantity.

It would seem, from the preceding experiments, that the stomach contains *no gastric juice*, in a free state, when aliment is not present. Any digestible or irritating substance, when applied to the internal coat, excites the action of the gastric vessels. Hence, I infer that the fluid, in these experiments, was incited to discharge itself by the irritation of the tube used in extracting it.

If, as is contended for by some, a *part* of the fluid be discharged into the stomach during a fast, I see no reason why nature should withhold the other part. If we may be allowed to argue, independent of more certain data, one great objection to the opinion that the stomach contains gastric juice, in a free state, when food is withholden from it, exists in the danger of its passing out through the pyloric orifice ; and thus depriving the succeeding meal of the benefit of its solvent action. It is probable that the pyloric orifice opposes no resistance to its egress ; but is obedient to its summons. In this way we may account for its admitting chyme, which is an admixture, or rather, combination, of gastric juice and food, to obey the expulsive motions of the stomach, and pass out. They both appear to excite the peculiar contraction of the pyloric end of the stomach, mentioned in a former part of this work. Besides, there would be danger of the gastric juice being weakened, by the introduction of large quantities of water, or other fluids, in the intervals of eating, and thus lose its energy, and concentrated solvent properties.

The last experiment has considerable pathological importance. In febrile diathesis, very little or no gastric juice is secreted. Hence, the importance of withholding food from the stomach in febrile complaints. It can afford no nourishment; but is actually a source of irritation to that organ, and, consequently, to the whole system. No solvent can be secreted under these circumstances; and food is as insoluble in the stomach, as lead would be under ordinary circumstances.

The following, and most of the subsequent experiments of this series, were instituted with the view of ascertaining the relative difference between *natural* and *artificial* digestion; to demonstrate the performance of digestion *out of the stomach*, by the gastric juice; and, also, the *continuation* of the natural process, when *taken out* during the period of chymification.

Experiment 24.

Dec. 14, 1829. At 1 o'clock, P. M., I took one and a half ounces of gastric juice, fresh from the stomach, after eighteen hours fasting, into an open mouthed vial—put into it twelve drachms *recently salted beef*, (boiled) and placed it in a basin of water, on a sand bath, and kept it at about 100° (Fahrenheit,) with frequent, gentle agitation. Digestion commenced, in a short time, on the surface of the meat, and progressed in that manner uniformly for about six hours, when its solvent action seemed to cease. The meat was at this time, nearly half dissolved; the texture of the central portion considerably loosened and tender, resembling the same kind of aliment when ejected, partly digested, from the stomach, some hours after being swallowed, as frequently seen in cases of indigestion.

The vial, continuing in the same situation, its contents varied considerable in their sensible qualities. In twenty-four hours, the digested portion separated into a reddish brown precipitate, and whey coloured fluid.

I now separated the undigested from the chymous portion, by filtration, through thin muslin. When squeezed dry, it weighed five drachms, two scruples, and eight grains, which, deducted from the twelve drachms of meat put in at first, leaves six drachms and twelve grains, digested in twelve fluid-drachms of gastric juice.

This experiment was conducted with as much precision and integrity of observation as possible, with the temperature of the digesting bath kept as near blood heat as was practicable to regulate and continue artificial warmth—the Thermometer varying, during the time, from 90° to 100°.

In this experiment, it appears, that it took *twelve* drachms of gastric juice to digest *six* drachms and *twelve* grains of aliment. No certain rule can, however, be given. Allowance must be made for the purity of the fluid, or its admixture with mucus and other secretions; for it is altogether probable that there are great variations in it, in this respect, as well as in all the other secretions of the system. It is probable, also, that different kinds of diet require different proportions of gastric juice for their solution. That its action is similar to that of other chemical agents I have no doubt. A given quantity of the fluid acts on a definite proportion of aliment, when it becomes saturated, and is inadequate to produce any further effect. There is always disturbance of the stomach when more food has been received than there is gastric juice to act upon it.

Experiment 25.

Dec. 16. At 2 o'clock, P. M.—twenty minutes after having eaten an ordinary dinner of *boiled, salted beef, bread, potatoes* and *turnips*, and drank a gill only of pure *water*, I took from his stomach, through the artificial opening, a gill of the contents, into an open mouthed vial. Digestion had evidently commenced, and was perceptibly progressing, at the time. This vial and contents were immediately placed in a basin of water, on the sand bath, at 90° or 100°, and continued there for five hours.

The digestion of the contents continued to progress, until all was completely chymified.

At 7 o'clock—five hours after eating his dinner—I took out a gill of pure chyme; no particles of undigested food appearing in the mixture.

Very little difference was perceptible between this last parcel and that in the vial, digesting on the bath. The stomach had digested a little faster and more perfectly than the vial.

In this experiment, it seems, that a quantity of aliment, taken out of the stomach *twenty minutes* after having been eaten, had a sufficient admixture of gastric juice to ensure its perfect digestion when placed on the bath. An ordinary moderate meal, taken into a healthy stomach, is sooner disposed of than most physiologists are aware of; and in this case, it is probable that a sufficient quantity of gastric juice had been secreted in twenty minutes, to digest the whole quantity of aliment in the stomach. When a larger quantity has been received, though the powers of the stomach may be sufficient, ultimately to dispose of it, it would undoubtedly be found that a portion taken from the stomach a few minutes after having been eaten, would not contain a sufficient quantity of gastric juice to digest it perfectly. It is possible that the portion presented at the perforation may be in a more advanced stage of digestion, than the rest of the mass, and consequently lighter, and float on the surface of the more solid portions of the food. In ordinary cases, such would be found to be the case; but when much fat meat or oily food has been used, the oil always maintains an ascendency in the gastric cavity.

Experiment 26.

Jan. 11, 1830. At 3 o'clock, P. M., dined on *bread* and *eight ounces* of *recently salted, lean beef, four ounces* of *potatoes*, and *four ounces* of *turnips*, boiled. In *fifteen minutes*, took out a portion of the contents of the stomach. The *meat* made its appearance, in an incipient stage of digestion.

At 3 o'clock, 45 minutes, took out another portion. The meat and bread only appeared, in a still more advanced stage of digestion.

The texture of the meat was, at this time, broken into small shreds, soft and pulpy, and the fluid containing it had become more opaque, and quite gruel-like, or rather, glutinous, in appearance.

I put this second parcel in a vial, and placed it in water, on the sand bath, at the temperature of the stomach, (100° Fahrenheit,) as indicated by the thermometer immediately preceding its extraction, and continued it there.

At 5 o'clock, took out another quantity. Digestion had advanced in about the same ratio as from the first to the second time of extracting; and when compared with the second parcel, contained in the vial on the bath, little or no difference could be perceived in them; both were nearly in the same stage of digestion. That contained in the vial had advanced regularly and rapidly; nearly all the particles of meat had disappeared, become chymified, and changed into a reddish brown sediment, suspended in the more fluid parts, with small particles, resembling loose, white coagulæ, floating about near the surface.

On taking out the third parcel, small pieces of vegetables appeared, in a partial stage of digestion. This was also put into a vial, and placed on a bath, with the second parcel, and the same uniform temperature (100°) kept up, with frequent, gentle agitation.

At 6 o'clock, P. M., digestion had progressed equally in both. The only difference to be seen, was the particles of vegetables, in a less advanced stage than the meat.

The contents of both vials, kept on the bath, and nearly in the same temperature, until the next morning, were completely digested, except the few small particles of vegetables, which remained almost entire.

The contents of the vials, at this time, were of the consistence of thin jelly, and of a lightish brown colour; tasting peculiarly insipid, saltish and acid. After standing at rest awhile, the brownish sediment subsided towards the bottom, while small particles of whitish coloured, loose coagulæ floated about in the fluid above. The undigested particles of vegetables settled to the bottom.

In the Preliminary Observations, I have endeavoured to maintain the proposition, that chyme is *homogeneous* in its properties. It would seem from this and some other experiments, that it contains a sediment. This, however, it is believed, does not militate against its homogeneous character. Many substances, that are generally acknowledged to possess this character, deposit a sediment, on standing. The heavier parts subside, of course. It is not necessary to cite examples. It is possible, also, that mixed food contains some adventitious, indigestible substances, which are not affected by the gastric juice.

This Experiment (26th) demonstrates the comparative digestibility of *animal* and *vegetable* diet. In extracting two parcels, one in fifteen minutes and the other in three quarters of an hour, the meat only made its appearance, partially digested. In taking out a third portion, at 5 o'clock, two hours after having eaten, small particles of vegetables made their appearance. This experiment appears to confirm the opinion, pretty generally entertained by medical men, that vegetables are less easily disposed of by the gastric organs, than animal or farinaceous substances. With dyspeptics this is undoubtedly true, as experience every day teaches us ; and as their stomachs have the same organization as others, are governed by the same general laws, and are only modified by debility or disease, the conclusion is natural, that they should act on aliment in the same manner, in proportion to their strength, that the healthy stomach does.

It may be inferred from this experiment, that the more perfectly chymified portions of food rise to the superior part of the stomach, as suggested in a previous observation, and are consequently exposed at the perforation, from whence parcels are taken for experiment and examination.

Experiment 27.

March 17. At 12 o'clock, M.—drank half a pint of *milk*. In fifteen minutes, took a portion out of the stomach, in a fine, loosely coagulated condition, per fectly white, and suspended in a semi-transparent, whey coloured fluid. I placed this on the bath, and it continued to digest for eight hours, when the coagulæ were completely taken up. A very small proportion of light coloured sediment, settled loosely to the bottom of a cream coloured, sweetish fluid.

At the same time that he drank the milk, I put one drachm of gastric juice, warm from the stomach, into two drachms of *milk*, and placed it on the bath, at the natural temperature, (100° Fahrenheit.) In five minutes, pure, white coagulæ formed, which, in fifteen minutes, exactly resembled that taken out of the stomach. In twenty minutes, the same fine, loose coagulæ were suspended in a similar liquid. These two drachms of milk, mixed with one drachm of pure gastric juice, *out of the stomach*, gave the same result, and exhibited the same appearance, in nearly the same time, as that which was *swallowed*, and *taken from the stomach*. Scarcely a shade of difference could be perceived in four hours.

Two drachms of *milk*, coagulated by *acetous acid*, produced coagulæ very similar to the other; but the wheyey part bore no resemblance, except in mere fluidity; that formed from the gastric fluid being of an opaline, slightly yellowish cast, and the other thin, transparent and watery.

The coagulæ formed by the gastric juice, continued to digest regularly in their fluid, for about eight hours, when they were completely taken up, and converted into chyme.

The coagulæ formed by the vinegar, remained in the same condition for forty eight-hours, with no other change except mere subsidence below the watery fluid.

It is well known, and this experiment was not necessary to prove it, that milk is coagulated before it receives the *solvent* action of the gastric juice. But it has some degree of importance in demonstrating the fact, that a degree of solidity is necessary for the operation of this agent. And it is a strong argument against the doctrine of digestion by the veins of the stomach. It has been maintained by some, that the veins take up the nutritious parts of the food, immediately on their introduction into the stomach. If so, it strikes me that they should do so, as it relates to this kind of aliment, while they are in a fluid state, and more susceptible of absorption by their mouths; and not wait till they have become solidified. Wine, spirits, water and other fluids, which conduce nothing towards alimentation, are neither coagulated, nor otherwise affected by the gastric juice. These fluids are not digested; and probably enter the circulatory system without much change.

It will be seen, by succeeding experiments, that other fluid, nutritive substances, particularly the albumen of eggs, are coagulated before they receive the solvent action of the gastric juice.

Experiment 28.

Jan. 25, 1831. At 1 o'clock, P. M., he ate a full dinner of *roast beef*, *potatoes*, *beets* and *bread*, and kept exercising about his usual employment, as house servant.

At 5 o'clock, 25 ms., I took out a portion of the contents of the stomach. Digestion of the different articles of food had commenced, and considerably advanced. The bread, reduced to a pultaceous condition, appeared floating about in a reddish brown fluid, of a glutinous consistence. A few small particles of the meat could also be seen in the fluid. None of the vegetables were discernible at this time. The fluids tasted slightly acid, giving the flavour peculiar to dilute muriatic acid, and very slightly bitter. A few grains of carbonate of soda, thrown into a drachm or two of this fluid, produced a slight effervescence.

At 4 o'clock, 20 mins.—took out another portion, a shade or two darker than the first. This dark colour of the chyme, I attributed to his having taken with his dinner, some of the outside, scorched pieces of the beef.

No distinct parts of the food could be seen at this time. Upon the surface of both parcels of fluids, floated a layer, of an oily or lardaceous consistence, which probably was the remains of the fat pork which he had eaten for his breakfast. The first parcel contained much more of this oily fluid than the last; which leads me to think that a considerable portion of an imperfect chyme, formed from the pork taken at about 10 o'clock, for breakfast, remained in his stomach when he ate his dinner; and then mixed with this aliment, in an imperfect state of digestion.

At 5 o'clock, 30 mins.--tried to extract another portion—could obtain nothing, except a little gastric juice. The chyme formed from his dinner appeared to have all passed from the stomach.

Experiment 29.

March 6. At 9 o'clock, A. M.—breakfasted on *venison steak*, *cranberry jelly* and *bread*, and drank a pint of *coffee*.

Twenty minutes after eating, I took a portion from the stomach, in an incipient stage of digestion. Placed this on the bath.

At 9 o'clock, 45 ms.–took out another portion, in an

advanced stage of digestion—very few small particles of food were discernible.

At 10 o'clock 10 mins.—took out another portion, completely chymified.

At 10 o'clock, 35 mins.—the stomach was entirely empty and clean—no chyme or aliment to be found in it. The breakfast, eaten at 9 o'clock, was all digested, and had passed through the pylorus, in *one hour and thirty-five minutes.*

This is an example of the great rapidity of digestion in some instances. This rapidity depends upon various circumstances—principally upon moderation in quantity, and the digestible properties of the food used. From various trials, I am confident, generally speaking, that venison is the most digestible of any diet of the fibrinous kind. In a few instances, it will be perceived, that other articles of diet were disposed of in a shorter period, than the venison was in this experiment.

Experiment 30.

March 7. Mixed two drachms of *albumen* of a fresh egg, with two drachms of gastric juice, warm from the stomach, and placed it on the bath, at the natural temperature. The juice and the albumen were so much alike in their appearance, when first mixed, that the change was not perceptible; but in ten or fifteen minutes, small, white flocculi began to appear, floating about; and the mixture became of an opaque and whitish appearance. This continued slowly and uniformly to increase, for three hours, at which time, the fluid had become of a milky appearance; the small flocculi, or loose coagulæ, had mostly disappeared, and a little light coloured sediment subsided to the bottom.

At the same time of the above experiment, he swallowed the *white* of two *eggs*, unmixed with any other food. The stomach was perfectly empty at the time.

In thirty minutes, I took out and examined a portion. It exhibited a similar appearance to that mixed out of the stomach, in the vial on the bath, only more rapid in its progress.

In *one hour and thirty minutes*, I examined the cavity of the stomach, and found nothing but a little pure gastric juice. The albumen was completely digested, and disposed of.

Experiment 31.

March 9. At 8 o'clock, A. M.—stomach empty—temperature 98°—took out two ounces gastric juice. Divided this into two equal parts, and put them into separate vials—to each of which, I put equal quantities of roasted beef—placed one of them on the bath, at 99°, and the other in the open air, at 34°.

I then put the same quantity of the meat into an equal quantity of clear water, and placed it with the cold gastric juice and meat in the open air, at the same temperature.

At 9 o'clock, he had finished breakfasting on the same kind of meat, with the addition of warm, light *biscuit*, *butter*, and a pint of *coffee*. Temperature of the stomach, immediately *before* eating, 100°. In thirty minutes *after* eating, the temperature rose to 102°.* Digestion rapidly advancing.

At 10 o'clock—took out a portion, partially digested; the biscuit the most so of any part of the breakfast. Placed this on the bath.

The meat, contained in the vial of gastric juice on the bath, was, at this time, in about the same condition as that taken from the stomach; very little dif-

* Probably the effect of exercise, but not noticed at the time.

ference could be perceived. The biscuit which he had eaten with his breakfast occasioned the only difference; that being reduced to a soft pulp.

The meat in the *cold* gastric juice was, at this time, much less advanced, than either that in the warm juice, or in the portion taken from the stomach.

That contained in the vial of water was merely macerated, and had no more appearance of digestion than what was effected by its being masticated, and mixed with the saliva, as were the other pieces of meat, before they were put into the gastric juice.

At 10 o'clock, 45 mins. I examined the stomach, but could find no distinct particles of food, and but very little chyme. His breakfast appeared to have been completely digested, and had left the stomach. Temperature 100°.

At 2 o'clock, P. M., the several parcels of meat placed in the gastric juice, on the bath, being about half digested, and appearing not to progress, I drew off twelve drachms of gastric juice from the empty stomach, and added four drachms to each, including the parcel taken from the stomach, at 10 o'clock, A. M., that being in about the same state of chymification with the others on the bath.

I continued the two on the bath, at 100°, and the others, (cold gastric and aqueous fluids) on the shelf, at 34°. Digestion evidently recommenced in the parcels on the bath, and again regularly progressed, after the addition of the second portions of the gastric juice; and more rapidly in the vial containing the meat digesting in the gastric juice, taken out of the stomach first, than in the one containing the chymous portion, taken out at 10 o'clock, A. M., one hour after having eaten. This parcel, however, contained a solid piece of meat, which appeared to have been swallowed without being masticated; and consequently did not readily yield to the solvent action of the gastric juice. The juice was, also, in too small proportion completely to digest it.

The vials containing the cold aqueous and gastric portions, placed on the shelf, were, at 4 o'clock, P. M., but very little changed, and much alike.

These four parcels, after standing for twenty-four hours, and all suffered to get cool, exhibited the following appearances.

The portion taken from the stomach at 10 o'clock, A. M., one hour after having eaten, was the most perfectly digested, and completely converted into a thick pultaceous mass, of a reddish brown colour, with the exception of the piece of unmasticated meat, which remained entire and undigested, This emitted a sharp, rancid smell, and was slightly bitter. The vial containing the meat digesting in the gastric juice first taken out of the stomach, exhibited appearances very similar to the last, though the contents were less perfectly digested. It was not of so thick consistence; but gave the same sharp smell and bitter taste, with the addition of an empyreumatic and slightly fœtid flavour. The empyreuma, I attributed to a portion of the meat being a little dry and scorched when first put in; and the fœtor, to the temperature of the bath having been accidentally raised considerably above 100°, during the experiment.

The cold gastric and aqueous portions very nearly resembled each other; both *macerated*, but not *digested;* differing essentially from the other two, in not exhibiting any appearance of chyme. The cold gastric juice had very little, if any, more effect on the meat, than the water; and retained its peculiar taste. Its colour was darkish brown, while the latter was of a reddish grey. At 9 o'clock, A. M., of the 10th, I placed both of them on the bath, and continued them for twenty-four hours, at the natural temperature. An essential difference in the gastric liquor was produced, after being placed on the bath. Digestion evidently advanced; the colour became lighter and lighter; the meat diminished; and a thin, light, paste-like liquor formed, as in the other two portions, at first placed on the bath. The aqueous portion exhibited no other appearance than that of simple maceration in warm water. At the end of the last twenty-four hours, on the bath, appearances of incipient putrefactive fermentation began to be manifested, as the evolution of small bubbles of fœtid gas, and a change of colour from a reddish to a greenish shade.

A difference in the degrees of chymification between the several parcels, was now very evident.

The gastric portion, or that taken from the stomach, an hour after breakfast, was the most digested.

The artificial, or that portion of the gastric juice and meat, first placed on the warm bath, was next, and nearly as much digested; though a difference was observable.

The third, or portion of gastric juice and meat, first placed in a cool situation, after having been on the warm bath for six or eight hours, was the next, but considerably less digested than the second.

The fourth, or aqueous portion, exhibited no appearance of chymification.

It would seem, from this experiment, that a certain degree of heat is necessary to the action of the gastric juice. One parcel of the meat, after being exposed to the cold gastric juice for twenty-four hours, exhibited very little change; but being placed on the bath, at the end of this time, digestion commenced, and advanced, regularly, as in the other parcels. It also appears, that after the process of digestion has ceased, for want of a sufficient quantity of gastric juice, it will recommence on the addition of a fresh supply. It was necessary to add another quantity, even to that portion taken out of the stomach, to ensure its perfect digestion. This, I think, is an evidence, that the fluid is discharged into the stomach gradually and progressively, according to the requirements of the aliment. If the portion left in the stomach had received, at the time the parcel was taken out, the whole quantity it was destined to receive, it must have been imperfectly digested, and have remained in the stomach, precisely in the situation of that which was taken out, and submitted to artificial digesiton; which is proved not to have received its full supply for perfect digestion. But subsequent examination demonstrated that it *was* perfectly digested, and had nearly all passed out of the stomach, in two hours. Hence, the conclusion is irresistible, that it received an additional quantity after the portion was taken from the stomach, one hour after eating.

Experiment 32.

March 12. At 8 o'clock, A. M., extracted one ounce of gastric juice.

At 9 o'clock, he breakfasted on *fat pork*, *bread* and *potatoes*. One hour afterwards, examined contents of stomach—found a heterogeneous mixture, resembling thick porridge.

At 1 o'clock, P. M.—four hours after having eaten—took out a portion, in a complete chymous state, without any entire particles of food to be seen. It was of a milky, or rather thin, gruel-like consistence, and considerably tinged with yellow bile; a circumstance which I had but once before observed in my experiments upon him. And this I supposed to have been the effect of violent anger, which occurred about the time of taking out this parcel.

This experiment shows the effect of violent passion on the digestive apparatus The presence of bile, in this instance, was believed to be the effect of anger. In a healthy state of the stomach, and an equable frame of mind, this substance has seldom been found in the stomach. When so found, except under peculiar circumstances of diet, it may generally be regarded as an indication of either mental or corporeal disease; and may be considered a foreign and offending substance in that organ. I believe its effect is to change the properties of chyme, (as it will be seen that it does, in subsequent experiments,) alter its homogeneous quality, and retard, or otherwise disturb its due egress into its destined receptacle, the duodenum.

Experiment 33.

March 13. At 1 o'clock, P. M.—dined on *roasted beef*, *bread* and *potatoes*. In half an hour, examined contents of stomach—found what he had eaten reduced to a mass, resembling thick porridge.

At 2 o'clock, examined again—nearly all chymified—a few distinct particles of food, still to be seen.

At 4 o'clock, 30 mins., chymification complete.

At 6 o'clock, examined stomach—found nothing but a little gastric juice, tinged with bile.

Experiment 34.

March 14. At 8 o'clock. 15 mins.—introduced two ounces of *rare*, *roasted beef*, suspended by a string, into the stomach; and at the same time, put one drachm of the same kind of meat into twelve drachms of gastric juice, contained in a vial, and put it into his bosom. The piece in his stomach, examined every hour, till 12 o'clock, M., exhibited an uniform, but very slow process of digestion, confined entirely to the surface of the meat. In four hours, about half of it, only, was dissolved and gone. That in the bosom, at the same time, digested still slower, owing, probably, to the circumstances, that the fluid in the vial had been taken out when the stomach was in a morbid condition, and had been permitted to get cold, even to the freezing point. This last circumstance, however, was probably, of less importance than the other. The meat in the stomach was too much confined by the string; was not permitted to move about freely in the gastric fluids by the natural motions of the stomach; and consequently did not digest so fast as it otherwise would have done Another circumstance or two, may also, have contributed to interrupt the progress of digestion, such as anger and impatience, which were manifested by the subject, during this experiment.

This experiment shows the necessity of a perfect comminution of the articles of diet. The gastric juice acted very slowly on a large, solid piece of meat. Digestion or solution was confined entirely to the outer surface. This, in addition to the other causes, mentioned above, produced the delay in digestion.

Experiment 35.

March 14. At 12 o'clock, M.—ate a *pint* of *milk*, and *four ounces* of *bread*. Examined stomach in thirty minutes—found the milk coagulated, and the bread reduced to a soft pulp, floating in a large proportion of fluid.

At 10 o'clock, 30 mins.—took out, and examined a portion—found it a thick pultaceous mass of bread, coagulæ and fluid, of a milky colour, slightly bitter taste, and acid smell. Placed it on the bath, where it continued to become more and more milky for an hour, when every particle seemed to be reduced to a rich fluid mass, resembling milk porridge.

The portion taken out thirty minutes after having been eaten, and kept on the bath, retained the appearance of the gastric fluid, with distinct flocculi of bread and coagulæ, floating about, and suspended in the fluid, and a little coarse precipitate at the bottom, after standing at rest a while.

At 2 o'clock—examined stomach—found it nearly empty. The bread and milk appeared to have been disposed of, and were gone from the stomach.

In this experiment, it took two hours for the digestion of a meal of bread and milk; something shorter than the usual time for the disposal of an ordinary meal. For those who have healthy and *unsophisticated* stomachs, milk appears to be one of the best articles of diet we possess. It is less stimulating than flesh, and more nutritious than vegetables. For persons who are disposed to pyrexial complaints, and who are not obliged to perform hard and exhausting labour, it is the most appropriate diet. But the stomach is a creature of habit. It can become accustomed to any kind of diet; and sudden changes are liable to derange its healthy actions. To those accustomed to what is called high living, such as strong meats, strong drinks, and high seasoned food, of all kinds, the transition to a milk diet, which contains a considerably lowered stimulation, would probably be an imprudent change. When necessary, the change should be so gradual, that the stomach should by degrees, become accommodated to it.

Experiment 36.

At 2 o'clock, 30 mins.–dined on *fresh beef and vegetable soup*, and *four ounces of bread.*

At 3 o'clock, 20 ms., examined contents of stomach —found a pulpous mass, of the consistence of thick gruel, and of a semi-gelatinous appearance. The soup appeared to have had its more fluid parts absorbed; for it was, at this time, much more consistent than when eaten. It was even thicker than the contents of the stomach usually are, after eating more solid food. Placed this on the bath.

At 5 o'clock, took out another portion, of a whitish colour, and more paste-like consistence, mixed with a little thin, transparent yellowish fluid, of an acid taste. The thick part had the flavour of bile, but not the colour.

Here the uniform laws with respect to liquid diet, appeared to govern the action of the gastric juice. The soup could not be digested until it was formed into a harder mass, by the absorption of the watery part. There was a less quantity of fluid than is usual after eating more solid food. This is another striking demonstration of the laws that govern the action of the stomachic solvent. If water were permitted to remain in the stomach, it would render the soup too liquid to be acted on by the gastric juice.

Experiment 37.

March 15. At 8 o'clock 30 mins. A. M.—breakfasted on *fresh sausage*, *light pancakes*, and a pint of *coffee.*

At 9 o'clock, 30 mins.—examined, and found the stomach full of fluids, mixed with the aliment; and a large portion of clear oil floated on the top, and presented itself at the perforation of the stomach.

At 10 o'clock, 30 ms., I took out a portion—found the cakes and particles of meat about half digested, with some oil, pure, bland and limpid, rising upon the top, untouched by digestion. Placed it on the bath.

At 12 o'clock, M., examined stomach—found no vestige of his breakfast—not a particle of oil was to be seen—nothing but pure gastric juice could be extracted, of which, I took out twelve drachms.

That portion of his breakfast, taken out at 10 o'clock and 30 minutes, was at this time, almost completely chymified, a few small particles of oil only remaining. The chymous mass of a milky colour, and thick, gruel-like consistence.

Experiment 38.

March 16. At 8 o'clock, 30 mins. A. M.—breakfasted on *fresh meat and vegetable hash*, *bread*, and a pint of *coffee*.

At 10 o'clock, 30 mins.—examined—found but very few particles of his breakfast in the stomach—some oil, and a few flocculi of a brown colour, run out with a little thin fluid.

At 11 o'clock—examined again—found nothing but a little gastric juice. Breakfast was gone, and the stomach clean.

These experiments, (37th and 38th,) are continued proofs of the solvent action of the gastric juice.

Experiment 39.

At 2 o'clock, P. M.—same day—dined on *recently salted*, *lean beef*, *pork*, *potatoes*, *carrots*, *turnips* and *bread*.

At 5 o'clock—examined—found the stomach clear of food, but containing a quantity of white, frothy mucus—villous coat inclined to dryness, and deeper pink colour. St. M. complained of some head ache, pain and distress at the pit of the stomach—dry skin and thirst. Directed him to take four drachms of *tincture of aloes and myrrh* at bed time. This operated two or three times next morning, and gave relief. The gastric juice, however, was not obtained in its usual quantity and quality, for twenty-four or thirty-six hours afterwards.

Experiment 40.

March 18. At 9 o'clock, A. M., he breakfasted on *soused tripe* and *pig's feet*, *bread* and *coffee*.

At 9 o'clock, 30 mins.—took out, and examined a portion—found it in a half digested condition, tripe, pig's feet and bread all reduced to a pulp, floating in a large proportion of fluids. Placed it on the bath.

At 10 o'clock—examined stomach again—tried to extract another portion—could find little or no chyme—a very little gastric juice, with a few small, fibrous particles of tripe, and some coffee grounds. His breakfast appeared to have been digested. and had passed from the stomach, in *one hour*.

The portion first taken out, and placed on the bath, was also, at the end of one hour, reduced to nearly a complete chymous condition; a very few of the small particles of tripe and coffee grounds only left, as in the stomach.

This is an example of astonishing rapidity of gastric solution; and that, too, of articles generally regarded as rather hard of digestion. That there could be no mistake, I infer from the fact, that a portion taken out of the stomach, thirty minutes after having been received, and submitted to the artificial mode, exhibited the same result.

Experiment 41.

At 1 o'clock, P. M.—same day—he ate eight ounces of *calf's foot jelly*, and nothing else.

In twenty minutes, examined stomach, and took out a portion of its contents, consisting of gastric juice, combined with the jelly, nearly all of it in a fluid form; a few particles only of entire jelly, suspended in the fluids, with a few small, yellowish white coagulæ, floating near the surface, could be perceived.

At 2 o'clock—examined again—extracted a little fluid, but found no appearance of jelly.

The operation of gastric juice on gelatine, is very difficult to be detected. Unlike albumen, it is unsusceptible of coagulation; and it is probable that the gastric juice acts upon it, in its soft-solid state. This was disposed of in a short period. It was, however, but a small quantity, and was much sooner digested than a full meal would have been.

From various trials, I am disposed to think that gelatine, if not in too concrete a state, is a very digestible article of diet.

During the examination, at this time, St. Martin swallowed part of a glass of water, and being situated in a strong light, favourable to an internal view, through the aperture, I distinctly saw the water pass into the cavity of the stomach, through the cardiac orifice—a circumstance, perhaps, never before witnessed, in a living subject. On taking repeated draughts of water, while in this position, it would gush out at the aperture, the instant it passed through the cardia. Food, swallowed in this position, could be distinctly seen to enter the stomach.

Experiment 42.

April 7. At 8 o'clock, A. M.—breakfasted on *three hard boiled eggs*, *pancakes* and *coffee*.

At 8 o'clock, 30 mins.—examined stomach—found a heterogeneous mixture of the several articles eaten, slightly digested.

At 8 o'clock, 45 mins.—examined again—found contents reduced in quantity, and changed in quality —about half digested.

At 10 o'clock, 15 mins., no part of the breakfast remained in the stomach.

This, and the four following experiments, throw no additional light on the subject of digestion, except so far as relates to the period of chymification. This, it will be perceived, depends something upon the *quantity* eaten. The quality, however, is not to be overlooked.

Experiment 43.

At 11 o'clock, 15 mins., A. M.—same day—he ate *two roasted eggs* and *three ripe apples.*

In thirty minutes, examined stomach—found a heterogeneous mixture, in an incipient stage of digestion.

At 12 o'clock, 15 mins., M.—examined again—found the stomach clear; no vestige of apples or eggs.

Experiment 44.

At 2 o'clock, P. M.—same day—dined on *roasted pig* and *vegetables.*

At 3 o'clock—examined, and found it about half chymified.

At 4 o'clock, very little remained in the stomach.

At 4 o'clock, 30 minutes, nothing remained but a very little gastric juice.

Experiment 45.

April 8. At 2 o'clock, P. M., he dined on *wild goose.*

At 3 o'clock—stomach full of fluids, with a large

portion of oil, floating on the surface; the goose flesh in small shreds, and soft; digesting rapidly.

At 4 o'clock—contents of stomach two thirds gone—that remaining, chymified.

At 4 o'clock, 30 mins., the stomach was empty and clean.

Experiment 46.

April 9. At 3 o'clock, P. M., he dined on *boiled, dried codfish, potatoes, parsnips, bread,* and *drawn butter.*

At 3 o'clock, 30 mins.—examined, and took out a portion, about half digested; the potatoes the least so of any part of the dinner. The fish was broken down into small filaments; the bread and parsnips were not to be distinguished.

At 4 o'clock—examined another portion. Digestion had regularly advanced. Very few particles of fish remained entire. Some of the potatoes were distinctly to be seen.

At 4 o'clock, 30 mins.—took out, and examined another portion—all completely chymified.

At 5 o'clock—stomach empty.

The preceding Experiments, I think, plainly demonstrate the *solvent* properties of the gastric juice. When aliment is submitted to it, *out of the stomach*, its operation is rather slower than when the process of digestion is assisted by the *natural warmth* and *motions* of that organ. One reason, probably is, the difficulty of maintaining a bath at the exact, necessary temperature; and another one may present itself, in the impossibility of perfectly imitating the motions of the stomach. With all these disadvantages, however, chyme formed in this way, presents the same uniform, sensible appearance, as that, which is formed in the stomach, by natural process.

That the *cold* gastric juice should not act at all, or but very imperfectly, on aliment, is no proof, in my opinion, that it does not possess solvent powers, even on the admission that it was a *debatable* question. There are but a few chemical combinations that do not require caloric to effect their operations, and none, perhaps, that are not facilitated by it. Some, and indeed, many of them require an intense heat. I am under the impression, though I have never fairly tested the truth of it, that gastric juice would, in a sufficient length of time, act on aliment, in a cold state. But I am not anxious to contend for any extraordinary or unnecessary powers of this fluid. Nor is it necessary to prove that it will act on cold substances, or in cold situations. It is perfectly manifest, that its operation is that of a chemical agent; that it dissolves aliment out of the stomach, when the warmth and motions of that organ are imitated; and that it changes the various and heterogeneous articles, submitted to its action, to an uniform, homogeneous semi-fluid, varying, however, slightly in colour and consistence, according to the aliment used.

With a view to ascertain, if practicable, what effects were produced by the BILE and PANCREATIC JUICE, when added to CHYME, I instituted the following Experiments.

Not being able to procure human bile, in a pure state, I obtained some *beef's gall*, and for pancreatic juice, substituted *diluted muriatic acid*, (one scruple acid to six ounces water.) I was induced to use this acid, from a resemblance observed between its taste and that of the pancreatic juice, and not being able to obtain any of that fluid at the time.

These experiments are certainly very imperfect but such as they are, I submit them to the public. They may tend to pave the way to more perfect experiments on these fluids.

Experiment 47.

I divided the chyme, produced in Experiment 24, Second Series, (Dec. 14th, 1829,) into two equal parts, about five drachms each. To one of which. I added one drachm of the Ox gall. Fine coagulæ were immediately produced, of a slightly yellowish green colour. To this, I then added one drachm of dilute muriatic acid; which immediately produced a white balsamic mixture. This, after standing at rest a few minutes, separated into three distinct parts; a clay coloured sediment at the bottom, a whey coloured fluid above, and a thin, oily, whitish pellicle on the top.

Experiment 48.

To an ounce of the chyme, formed in Experiment 25, (Dec. 16th,) I added one drachm of the Ox gall; which immediately converted it into a milky fluid, very finely coagulated. To this, I added one drachm of the diluted muriatic acid, which at first, increased the coagulæ; but immediately after, threw down a brown precipitate. This, on the addition of more bile and acid, varied in colour, according to the different proportions put in, from a light clay colour, to a dark brown, tinged with green, without any change in the colour or consistence of the fluid above.

On standing at rest, it separated into three distinct parts, a brown sediment at the bottom, a yellowish or whey coloured fluid in the middle, and a thin, milky white pellicle on the top.

Experiment 49.

Having procured some fresh gall, from an Ox recently slaughtered, I added twenty drops of it to four drachms of the chyme formed in Experiment 26, (Jan. 11th, 1830.) A turbid, yellowish white fluid, or rather, very fine, cream-coloured coagulæ, immediately formed; which, after standing a few minutes, separated into bright, yellow coloured coagulæ, subsiding towards the bottom, and a turbid, milk coloured liquid above.

By adding twenty drops more of the bile to this, the coagulæ were increased, more collected together, and changed in colour, from a yellow to a greenish hue.

The addition of twenty drops more of bile, (making, in the whole, one drachm,) concentrated a deep grass green, jelly-like deposition at the bottom of the vial. The fluid above, became more milky in appearance; and the coagulæ and sediment became darker on the addition of bile.

I now added twenty drops of the dilute muriatic acid to other four drachms of the same kind of chyme, without bile. This produced no change in the colour or consistence, but increased the saline, acid taste, peculiar to the gastric and pancreatic juices, when uncombined with chyme.

By adding bile to this, the same effects and appearances were present as in the other similar experiments; viz.: a yellowish brown sediment at the bottom, a whey coloured fluid in the middle, and a white pellicle on the top.

To observe the different effects produced between a combination of bile and muriatic acid in clear water, and that of the chymous mass, I mixed equal quantities of the gall and dilute acid, one drachm each, with two ounces of water. This at first produced an effect, and exhibited an appearance, similar to that of their combination with chyme; but gradually changed to a bluish, green coloured, thin fluid, with a deep green, jelly-like deposition at the bottom, without any of the milky appearance of the chymous mixtures, or white pellicle on the top.

Experiment 50.

To four drachms of gastric juice, fresh from the stomach, I added forty drops of Ox gall, which produced a turbid, yellowish green fluid, yielding no sediment.

Forty drops dilute muriatic acid, added to other four drachms of the gastric juice, effected no change in its appearance.

Equal parts of the bile and muriatic acid, mixed together, produced a fluid of exactly the same colour as the first; but was less consistent.

On mixing the two first together, and adding two drachms of chyme from the stomach, very fine coagulæ formed in a milky fluid, throwing down a brownish sediment, from a whey coloured liquor, with the same milky pellicle on the surface, as in the former experiments.

To one ounce of chyme, formed in a vial, on the bath, I added two drachms of bile. A turbid, yellowish white mixture formed, without sediment, or immediate separation of any kind.

To another ounce of the same chyme, I added two drachms of the dilute acid. No change in its appearance was perceptible.

I then mixed them together, and the appearance of both was changed. Whitish coagulæ formed, and let fall a brown sediment, leaving an opaque, whey coloured fluid above, with a pellicle or white flocculi on the surface.

Experiment 51.

Bile added to the third portion of chyme, taken from the stomach one hour and ten minutes after a breakfast of venison steak, &c., Experiment 29, (March 6th, 1831,) changed it from a brownish, homogeneous paste, to a milky fluid, with small, white flocculi, floating about, or adhering to the sides of the vial: and a light brown sediment settled to the bottom.

The usual proportion of dilute muriatic acid, added to this, produced no very essential change in its appearance, causing only a little more deposition of sediment, and slightly increasing the milky colour.

Experiment 52.

Bile added to the chyme formed from the eggs, digested out of the stomach, Experiment 30, (March 7th, 1831,) produced a rich, milky fluid, with a small quantity of fine, light coloured sediment, falling to the bottom.

The dilute acid, added to this, produced fine coagulæ, and formed a milk white whey, or fluid, from which, more of the light coloured sediment was precipitated.

Experiment 53.

More minutely to observe the respective changes by the addition of bile and muriatic acid, in the several parcels of chyme formed in Experiment 31, (March 9th, 1831,) and to note their difference, I put equal quantities of each into glasses, and added a portion of hog's gall.

In the first, (that taken from the stomach at 10 o'clock, one hour after having eaten,) fine, bright orange coloured coagulæ were immediately formed, equally diffused through a fluid of the same colour, exhibiting no perceptible sediment on standing at rest; but held the coagulæ, uniformly suspended throughout the fluid. The dilute acid, added to this, occasioned a copious sediment to fall to the bottom, and with it, all the colour of the mixture, leaving a transparent, semi-gelatinous-like fluid above, in the proportion of about three-fifths of the whole; upon the surface of which, floated a thin, white pellicle.

The second portion, (that produced on the bath) under the same treatment, exhibited nearly the same appearance, with the exception of the colour, which was a shade or two lighter. The sediment was not quite so compact; the fluid less gelatinous; and there was less of the white pellicle on the surface.

The third portion, treated like the other two, differed about as much from the second, as this did from the first. They all exhibited the same general appearance.

The fourth, or aqueous portion, under the same treatment, exhibited a wide difference. The same proportion of bile added to this, produced a similar coloured fluid, at first, with a very little coarse coagulæ—not so uniformly diffused through the liquid; but inclining more to precipitation. On adding the acid, it let fall a very small quantity of yellowish green sediment, leaving a thin, semi-transparent fluid, in more than quadruple the proportion of the other three.

Experiment 54.

Bile and dilute muriatic acid, added to a portion of the bread and milk chyme, formed in Experiment 35, (March 14th,) produced their usual coagulation and precipitation, but of a lighter yellow; the sediment forming about one fourth of the mass. The small, white particles, forming the pellicle on the top, were in greater proportion than in some of the other experiments, especially those on lean meats. The fluid part was in greater proportion to the sediment, and of a whey colour and consistence.

To another equal quantity of this same kind of chyme, I added bile, as in the other, and instead of muriatic acid, I used *pancreatic juice*, fresh from a recently slaughtered beef. An appearance exactly similar to that produced by the acid, was exhibited, except that the precipitate was more slowly thrown down, and in larger proportion; and the white pellicle on the surface was less. The fluid and sediment were a shade lighter, and in more equal proportions.

Experiment 55.

Pancreatic juice, combined with the chyme of roast beef, formed both in and out of the stomach, increased its thin, paste-like consistence, and gave it more of a cream colour. Bile, added to this, produced fine coagulæ, suspended from the top to the bottom, without depositing any distinct sediment. Diluted muriatic acid darkened the whitish colour, a shade or two, threw down a more copious sediment, and increased the white pellicle on the top.

Experiment 56.

Bile and pancreatic juice, added to the fresh meat and vegetable soup chyme, Experiment 36, (March 14th, 1832,) produced loose, cream coloured coagulæ; which, on standing, separated into three, about equal proportions; a coarse, brownish sediment, a semi-transparent, whey coloured fluid, and a thick, white pellicle at the top.

EXPERIMENTS, &c.

THIRD SERIES.

WASHINGTON, D. C. 1832.

Experiment 1.

Dec. 4. At 2 o'ck, 30 ms., P. M.—Weather cloudy, damp and snowing—Th : 35°—Wind N. W. and brisk—the temperature under the tongue was 99° ; in the stomach, 101°. Dined, at 3 o'clock, 30 mins., on *beef soup*, *meat* and *bread*. 4 o'clock, 15 mins.—took out a portion—particles of beef slightly macerated, and partially digested. 5 o'clock, 15 mins.—took out another portion—digestion more advanced—meat reduced to a pulp ; particles of bread and oil floating on the top. Temperature of stomach, 100°. 6 o'clock, 45 mins.---digestion not completed—contents considerably diminished. 7 o'clock, 45 mins.—stomach empty—chyme all passed out.

Experiment 2.

Dec. 5. At 7 o'clock, A. M., temperature of the stomach, 100° ; of the atmosphere, 30°.

At 1 o'clock, P. M.—temperature of stomach, 100° —atmosphere, 40°—he ate eleven *raw oysters*, and three *dry crackers ;* and I suspended one *raw oyster* into the stomach, through the aperture, by a string. 1 o'clock, 30 mins.—examined – stomach full of fluids---digestion not much advanced. The oyster on the string appeared entire, though perhaps slightly affected on the surface. 2 o'clock—examined, and took out oyster--about one third digested, but retained its shape. 2 o'clock, 30 mins.---oyster gone from the string, except a small piece of the heart. Temperature of the stomach 101½°. Fluids less considerable. 4 o'clock, 15 mins.--stomach empty.

Experiment 3.

At 3 o'ck, 45 ms., P M., same day, he dined on *roast turkey, potatoes* and *bread.* 4 o'clock, 30 mins.---examined, and took out a portion. Turkey nearly all dissolved—vegetables half reduced. 5 o'clock, 15 mins.---took out another portion, almost completely chymified. 5 o'clock, 45 mins.---examined again—stomach nearly empty. 6 o'clock---some chyme yet remaining. 6 o'clock 15 mins.---stomach empty.

Experiment 4.

Dec. 6. At 8 o'clock, 30 mins., A. M., he breakfasted on *bread and butter*, and one pint of *coffee.* 9 o'clock, 45 mins.---examined—stomach full of fluids. 10 o'clock, 30 mins.---examined, and took out a portion, resembling thin gruel, in colour and consistence, with the oil of the butter floating on the top; a few small particles of the bread, and some mucus, falling to the bottom.--about two thirds digested. It had a sharp, acid taste. Temperature of the stomach, 100°---atmosphere, 38°. 11 o'clock, 30 mins., stomach empty.

Experiment 5.

At 4 o'clock, 30 mins., P. M., same day--he dined on *sausage* and *bread ;* full meal. 5 o'clock, 30 mins. ---stomach full of fluids ; digestion but very little advanced. 6 o'clock, 30 mins.--digestion considerably advanced : few distinct particles of sausage and bread to be seen entire. 7 o'clock, 30 mins., stomach empty.

Experiment 6.

Dec. 7. At 8 o'clock, A. M.,---examined stomach, and took out, with considerable difficulty, an ounce only, of gastric juice, and that not very pure. Some yellow bile came mixed with the latter portions. Temperature of the stomach, 99°--atmosphere 28°. He breakfasted, at 9 o'clock, on *corn and wheat bread, butter*, and *coffee.*

At 10 o'clock, 45 mins.---examined, and took out a portion--food partly digested; few small particles to be seen. Stomach full of fluids, with a thin pellicle of oil on the top. Temperature of the stomach, 100°.

At 12 o'clock, M.--stomach full of fluids---digestion not complete--particles of bread floating about in a pulpous state---oil floating on the surface.

At 12 o'clock, 30 mins., M.,--examined--contents of stomach half diminished--distinct particles of oil on the surface.

At 12 o'clock, 45 mins.---entire particles of bread, yet to be seen---quantity of fluid diminishing.

At 1 o'clock, P. M.—distinct particles of bread still floating---fluid less.

At 1 o'clock, 15 mins.---stomach empty.

Some indications of gastric derangement this morning: small aphthous patches on the mucous membrane: juice acrid and sharp, with bile mixed with it.

Experiment 7.

At 3 o'clock, 30 mins., P. M., same day, he dined on *roasted mutton, bread* and *potatoes.* 4 o'clock, 45 mins. ---examined---stomach full---digestion advancing. 5 o'clock, 45 mins.---contents of stomach three quarters reduced in quantity, and almost completely chymified. 6 o'clock, 30 mins.---stomach nearly empty; a little pulp of the bread only to be seen, floating in a little milky fluid. 7 o'clock....stomach empty.

Experiment 8.

Dec. 8. At 5 o'clock, 30 mins., A. M....temperature of stomach, 99°. 9 o'clock....finished breakfasting on *fried sausage*, *dry toast*, and a pint of *coffee.* 10 o'clock, 30 mins....stomach full of fluids....villous coat red and irritable, inclining to dryness....a thin, whitish coat on the tongue, and a similar appearance on the protruded portion of the stomach. 11 o'clock, 45 mins. ....stomach full....oil floating on the top, and rancid. Temperature of stomach, 99°....atmosphere, 46°. Weather damp and cloudy.

This, and the 6th Experiment, show, that when there are indications of disease on the coats of the stomach, and on the tongue, digestion is consequently protracted; and, also, that oil is particularly hard of digestion.

Experiment 9.

At 9 o'clock, A. M., same day, the vial containing the *bread and butter* aliment, taken from the stomach on the 5th inst. (Experiment 4,) at half past 10 o'-clock, A. M., was placed on the bath for four hours, in the usual temperature, between 95° and 100°. Digestion commenced, and advanced regularly, partially reducing the oil to a milky fluid.

Dec. 9. At 11 o'clock, A. M....added one ounce of gastric juice, and continued it on the bath for eight hours, when the oil became more, but not completely digested; particles of the limpid oil being still perceptible.

This affords an example of the re-commencement of digestion, after the operation had ceased, by the addition of a fresh supply of gastric juice.

Experiment 10.

At 2 o'clock, 45 mins., P. M., same day, (*Dec.* 8,) I suspended a *roasted oyster*, weighing, when raw, four drachms, into the stomach, and he ate twelve of the same kind, each weighing about the same.

At 4 o'clock, 30 mins....examined....oyster remaining on the string, not half digested....fluid in the stomach rancid. Complained of head ache, lassitude, dull pains in the left side, and across the breast....tongue furred, with a thin, yellowish coat, and inclined to dryness....eyes heavy, and countenance sallow. The villous membrane of the protruded portions of the stomach, very much resembled the appearance of the tongue, with small aphthous patches, in several places, quite irritable and tender.

I suspended observations, and dropped into the aperture at night, six grains *blue pill*, and four *aloetic pills*, (common size,) and sprinkled on the exposed surface of the stomach, five or six grains of *calomel*. Medicine operated early the next morning; relieved the symptoms of indisposition; changed the appearance of the stomach and tongue; and removed the aphthæ. On the 9th, he felt quite well; and the coats of the stomach looked healthy again.

Experiment 11.

Dec. 13. At 7 o'clock, A. M....temperature 100° ....villous membrane perfectly healthy, of a pale pink colour, and uniform....mucous coat smooth and even. Extracted two ounces of gastric juice. It distilled more freely than common. More could have been obtained. I had never before seen the pure juice flow so freely. He felt in perfect health: had taken neither food or drinks since 9 o'clock, last evening.

At 9 o'clock—breakfasted on *broiled breast of mutton*, *bread*, *butter*, in usual quantity, and a pint of *coffee*, and kept exercising. Digested in three hours and a half; stomach empty and clean.

Experiment 12.

At 2 o'ck, P.M., same day—stomach empty—coats clean—he dined on three *soft boiled eggs* and *bread*, and drank half a pint of *water*. 3 o'clock— digestion advancing. 4 o'clock—contents nearly gone from the stomach—yolk of eggs still visible, with a few particles of oil. 5 o'clock—very little chyme in the stomach. 5 o'clock, 15 mins.—some still remaining. Complains of slight head ache—pulse full and crowded—contents of stomach acrid—countenance rather sallow; eyes languid; tongue a little coated with a thin, yellowish fur. His bowels have not been moved since yesterday morning, at 10 o'clock; then inclined to costiveness.

N. B. After taking breakfast, he exercised moderately. About 12 o'clock, M., he walked about two miles very quick. After his return to his lodgings, he threw off his coat, and went into the open air again. Soon after which, he began to feel the pain in his head, &c.

Experiment 13.

Dec. At 7 o'clock, A. M.—stomach deeper colour than ordinary, and inclined to dryness—some small, aphthous patches, and spots of darker colour—mucous coat not uniform and even; some places thicker, a little elevated, and rolling up, like thin membrane, leaving a spot beneath, red and irritable. Very little juice could be extracted. I obtained a small quantity of fluid, mixed with yellow bile. It did not yield the peculiar acid taste of the gastric juice. Temperature of the stomach, 100°. St. Martin did not feel his usual appetite.

At 9 o'clock, he breakfasted on the same kind of diet as yesterday—had less appetite, and was labouring under some gastric derangement. He continued quiet, most of the time in a recumbent position. 10 o'clock—stomach full—globules of oil floating about —appearance of villous membrane, about the same; no perceptible change. 11 o'clock—stomach still full—appearances similar to those in last examination. 12 o'clock, M.—contents half diminished—

particles of bread, and coat of oil on the surface. 1 o'clock, P. M.—some fluid still in the stomach, and a larger proportion of oil than at last examination. Taste of the contents, more sharp and rancid; fast leaving the stomach. At this time, I observed several small, sharp pointed, white pustules or pimples, here and there dispersed over the exposed portion of the inner coat. 1 o'clock, 30 mins.—stomach clear and clean.

Experiment 14.

At 2 o'clock, P. M., same day, he dined on three *soft boiled eggs*, *bread and butter*, and half a pint of *water*, (same as yesterday, 2 o'clock.) Digested in three hours.

Experiment 15.

Dec. 15. At 8 o'clock, A. M., I examined stomach —temperature 100°. Appearance of coats more natural and healthy than yesterday morning; less of those small, white, pointed pimples, and aphthous spots. Very little gastric juice could be obtained; not more than one ounce, and that mixed with an unusual quantity of mucus, not so clear as common. Complained, as he frequently does, during this operation, of a sense of sinking, and vertigo, after extracting this quantity. This feeling, however, subsided in a few minutes after rising.

At 8 o'clock, 30 mins. he breakfasted on *beef steak*, *bread* and *coffee*. At the same time, he thoroughly masticated four drachms of the steak, which I put into the gastric juice, just before taken from the stomach. To another similar quantity of gastric juice, I put the same quantity of the steak, unmasticated, and in one entire piece. I placed them both on the bath at 100°; and at the same time, I put the same quantity steak into one ounce of simple water, and treated it with the others on the bath.

At 11 o'clock, I examined the stomach, and found his breakfast nearly digested, and more than half gone from the stomach. I took out an ounce of what remained, which was almost completely chymified, a few particles of the bread, in a soft, pultaceous condition, only remaining. Compared this with the three parcels on the bath. It very nearly resembled the masticated meat in the gastric juice, but more digested, and thinner, and contained particles of oil (melted butter) and bread, which were not in the masticated food in the vial. The unmasticated meat differed considerably. It was not so thick and gelatinous-like; was of a darker colour; and the piece of meat retained its shape, and was not much diminished in size, the surface only a little wasted, softened, and covered with a cineritious coat.—The contents of the vial of masticated meat and water, suffered very little or no change since put in; no more than had been effected simply by mastication. Continued them all on the bath.

The contents of the vials, continued on the bath for twenty-four hours, exhibited the following changes. The portion taken from the stomach at 11 o'clock, remained nearly the same as when extracted, perhaps more completely chymified. That which was masticated, and put into the gastric juice, was reduced to a thick, pultaceous, semi-fluid mass, but retaining some distinct fibres of the meat, which, after standing awhile, subsided to the bottom of a yellowish, whey coloured fluid. These remaining particles of aliment, I conceived to have been left for want of a sufficient quantity of gastric juice; the quantity at first being too small to dissolve the whole of the meat put in. That portion in the vial of water had undergone no other change than that of incipient putrefaction, which was very evident. The unmasticated piece of meat had undergone an evident process of digestion. It was about half diminished, and the texture of the remaining part loose and soft. The containing fluid had become of a greyish-brown colour, opaque, with a fine, brown sediment, settling to the bottom, similar to that of the masticated meat in the gastric juice. The gastric juice, containing the unmasticated meat, when taken from the stomach, some sixty or seventy hours before, was not so pure as common; was mixed with yellow bile; and was in too small proportion to the meat. The colour and flavour of the other two portions were very similar, except that the one with the masticated meat was more sharp and acrid.

This experiment shows the necessity of mastication; and also demonstrates, that simple maceration, at the natural temperature, will not effect digestion.

Experiment 16.

A dinner of *pork steak* and *bread*, taken at 1 o'clock, P. M., same day—digested in three hours, forty-five minutes.

Experiment 17.

Dec. 16. At 9 o'clock, A. M., he breakfasted on *cold, pork steak, bread*, and one pint *coffee*. Digestion completed in three hours. Two hours after having eaten, a pellicle of oil was found floating on the top of the gastric contents.

On examining the stomach, an hour after the chyme had passed out, several red spots and patches, abraded of the mucous coat, tender and irritable, appeared spread over the inner surface. The tongue, too, had upon it a thin, whitish fur. Yet his appetite was rather craving.

At 2 o'clock, 30 mins., P. M., he ate a full dinner of *cold, roasted pork*, (fresh) *bread*, and a piece of *raw radish*. Digestion completed in seven hours.

Experiment 18.

Dec. 17. At 8 o'clock, 30 mins., A. M., I put two drachms *fresh, fried sausage* in a fine muslin bag, and suspended it into the stomach. He immediately after breakfasted on the same kind of *sausage*, and a small piece of *broiled mutton*, *wheat bread*, and a pint of *coffee*. 11 o'clock, 30 mins., stomach half empty —contents of bag about half diminished. 2 o'clock, P. M., stomach empty and clean—contents of bag all gone, except fifteen grains, consisting of small pieces of cartilaginous and membranous fibres, and the spice of the sausage; which last weighed six grains; leaving only nine grains of the aliment put in. In consequence of being called out, I delayed the last examination longer than was necessary.

Experiment 19.

Dec. 18. At 8 o'clock, 30 mins., A. M., I suspended two drachms masticated, *fried sausage*, confined in a muslin bag, into the stomach, and he breakfasted on the same kind of food, with *bread* and *coffee*. 11 o'clock, 30 mins., stomach half empty—contents of bag about half gone. 1 o'clock, P. M., stomach nearly empty—very little left in the bag. 1 o'clock, 30 mins., stomach clear, except the bag, which contained a little of the sausage: took this out, and it weighed one drachm, spice and all, of which there was less than yesterday. The bag, when drawn out, came from near the pylorus, and was covered with a coat of mucus and yellow bile. The contents of the stomach have been unusually acrid since yesterday morning, and he complains of unusual smarting and irritation at the edges of the aperture: countenance sallow; tongue covered with a thin, yellowish coat; and several deep red patches on the inner coat of the stomach: does not feel his usual appetite. 9 o'clock—dropped into the aperture, twelve grains *blue pill*, and five *cathartic pills*—operated early the next morning; removed the symptoms; and restored his healthy sensations and functions.

Experiment 20.

Dec. 19, At 8 o'clock, 45 mins., A. M., I suspended three drachms *broiled bass*, in a muslin bag, into the stomach, and he breakfasted on the same kind of fish, with *bread*, a small piece of *sausage*, and a pint of *coffee*. 2 o'clock, P. M.—complains of smarting at the aperture--I took out the bag—remaining contents weighed two drachms, having lost one drachm only in five hours and a quarter. Coats of the stomach did not appear healthy—deeper red than natural, with patches of still deeper colour, spread over the protruded portion. Mucous covering abraded in places, and rolled up; resembling shreds of epidermis, torn from a blistered surface.

These three last experiments, are examples of the solvent or chemical action of the gastric juice. It penetrated the muslin bags, dissolved the food, and allowed the chyme to strain out. They also indicate that irritating substances, (as, for instance, the muslin bags, in these experiments,) produce a diseased state of the stomach.

Experiment 21.

Dec. 20. At 8 o'clock, 30 mins., A. M.—Coats of stomach appear healthy—considerable fluid plainly to be seen. It ran out of the aperture on turning him down; was transparent, and contained flocculi of mucus. Breakfasted on *broiled bass*, *toasted bread* and *coffee*. Digested in five and a half hours.

Experiment 22.

At 2 o'clock, P. M., he dined on *boiled chicken*, and *wheat bread*. Digested in four and a half hours.

Experiment 23.

Dec. 21. At 8 o'clock, 30 mins., A. M.—stomach not perfectly healthy—several small, deep red patches, on the exposed surface. Extracted four drachms gastric juice, tinged with yellow bile. Masticated one and a half scruples of the thigh of a *boiled chicken*, and half a scruple of bread: put them into this gastric juice, and placed the vial in the axilla. Into the same quantity of pure water, warmed to 70°, I put the same quantity and kind of aliment, and placed them in the same situation. He breakfasted at the same time, on the same kind of diet. 1 o'clock, P. M.—stomach empty. At 2 o'clock, he dined on same kind of food. 6 o'clock, 30 mins.—stomach empty.

The masticated portion put into the vial of gastric juice, placed on the bath, and frequently agitated, digested regularly and uniformly until about 2 o'clock, P. M., when the particles were all dissolved, except a few fibres. That in the vial of water, kept in the same situation, had not changed its appearance from the time it was put in.

On separating the remaining particles of food, in the gastric juice, at evening, filtering on thin muslin, and drying with paper, it weighed fifteen grains, and left four drachms and a fraction, of an opaque, milky coloured fluid.

That in the water, taken out at the same time, weighed forty grains, and left four drachms of a turbid fluid, like water, with flour stirred in it, and had a mawkish, insipid taste and smell. The first had the acid smell and taste, peculiar to the gastric contents.

Experiment 24.

Dec. 22. At 8 o'clock, A. M.—examined stomach —temperature 100°. Extracted about four drachms gastric juice, pure, but not free.

At 8 o'clock, 30 mins., he breakfasted on *bread cheese* and *coffee.* 9 o'clock, stomach full of fluids—temperature 100°. 11 o'clock—stomach full, with the cheese in a fluid form, floating on the surface; bread reduced to a pulp—temperature 100°. 12 o'clock, M.—food still in the stomach; but considerably diminished. 1 o'clock, 30 mins., P. M.—some of the cheese yet remaining—stomach nearly empty. 2 o'clock—stomach empty.

The coats of the stomach have not appeared in their usual healthy condition, for several days past —the colour darker—mucous coat unequal—some patches of a purplish colour, with aphthous edges—surface inclined to be dry—very little secretion of gastric juice—digestion slower, and less perfect than usual--bowels inactive, nothing having passed them for sixty hours.

It would seem from this experiment, that cheese was difficult of digestion. In addition to its closeness of texture, it generally contains a large proportion of oil.

Experiment 25.

Dec. 23. At 6 o'clock, A. M.—temperature of stomach, 100°—pulse 65 a minute. 9 o'clock—temperature of stomach, 100°—pulse 75. Villous membrane inclined to dryness, and of a darker than natural colour; papillæ small and sharp; mucous covering scarcely perceptible; bowels costive; tongue coated with a yellowish fur, and its edges pale. I poured in, at the aperture, one ounce *Ol. Ricini*, and sprinkled over the surface of the protruded coats, five or six grains of *calomel*.. He ate a light breakfast of *corn bread* and *crackers*, and drank a pint of *coffee*, immediately after.

At 2 o'clock, P, M.—stomach empty—coats look healthier. Medicine not having moved the bowels, I put in, at the aperture, twelve additional grains of *calomel*, per se.

At 5 o'clock, the stomach was in commotion—indications of the cathartic operation of the calomel: slight nausea; stomach full of a white, frothy fluid, running out at the aperture, like fermenting beer from a bottle; slight pain and motion in the bowels; and increased secretion of saliva. No motion from the bowels. Temperature of stomach, 101°. Pulse 80 beats in a minute.

At 8 o'clock, calomel had operated twice, copiously, commencing at 7. Temperature of stomach, 100°. Pulse 62, soft and mild.

Experiment 26.

Dec. 25. At 8 o'clock, A. M.—weather partially cloudy—atmosphere dry, and smoky—wind E. and light—Th: 31°. Temperature of the stomach, 100° and a fraction. Pulse 55, in a recumbent position; 65, sitting erect. A few small, red spots, on the mucous surface. The gastric secretions appear as healthy as usual.

At 9 o'clock, he breakfasted on *boiled, salted, fat pork, corn bread* and *coffee.* 10 o'clock, the stomach at the same temperature as at 8 o'clock. Pulse 65 in a recumbent, and 75 in an erect position. Gastric cavity full of a heterogeneous mixture.

At 11 o'clock, 30 mins.—just returned from walking moderately, about an hour, a distance of two and a half miles; not to produce free perspiration, but gentle diaphoresis. Weather clear, calm and dry. Th: 50°. Temperature of the stomach 101°. Pulse 72, in a recumbent position; 82, sitting erect, and regular. Contents of stomach half reduced, and nearly homogeneous.

At 12 o'clock, 30 mins., M.—temperature of stomach, 100½°. Pulse 62, recumbent; 72 erect. Contents nearly gone.

At 1 o'clock, 30 mins., P. M., stomach empty.

At 9 o'clock,—weather cloudy—atmosphere dry—no wind—Th: 42°---the temperature of the stomach was 99½°. He drank half a pint of water fifteen or twenty minutes before examination. Pulse 62, recumbent; 72, erect.

This is an example of the increase of the temperature of the stomach on exercise. See also, subsequent experiments.

Experiment 27.

Dec. 26. At 6 o'clock, A. M.---weather cloudy; atmosphere damp; wind N. E. and light; Th: 38°; temperature of the stomach, 99½°. Pulse 55, recumbent; 65 erect. Respirations, in a recumbent position, 15, and in a sitting position, 18 a minute.

At 8 o'clock, he returned from a walk of two miles, but not to produce perspiration. Weather damp and raining lightly. Th: 36° Temperature of the stomach, 101°. Pulse 65 recumbent; 85, erect. Feelings of impatience here evidently accelerated his pulse, in the erect position. He was vexed at being detained a few minutes from his breakfast.

At 5 o'clock P. M.--- weather rainy---wind N. E.... Th: 41°-- I examined the stomach. Temperature, 99½°. Pulse 60, recumbent; 70, erect.

At 8 o'clock, the temperature of the stomach, 101°. Pulse 50, recumbent; 60, erect. Respirations, 15 a minute.

His diet through the day had been confined principally to farinaceous substances, wheat bread and crackers, in moderate quantities.

Experiment 28.

Dec. 27. At 6 o'clock, A. M. Weather unpleasant. Atmosphere damp. Wind E. Th: 38°. Temperature of stomach, 99½°. Surface clean and healthy. No dark red, or aphthous patches, nor white, elevated points. Mucous coat uniform and even, of the natural colour. No excoriation or smarting at the edges of the aperture. I extracted one ounce of gastric juice, slightly tinged with yellow bile. This, I conceive to have been entirely accidental; and occasioned by the regurgitation of the bile through the pylorus, as he turned upon his back, from right to left, to favour the exit of the gastric juice. The same thing has happened several times before.

At 9 o'clock, he breakfasted on three ounces *broiled breast of mutton*, four ounces of *wheat and corn bread*, very thoroughly masticated, and a pint of *coffee.* At the same time, I put two drachms of same kind of food, equally well masticated, into the ounce of gastric juice, taken from the stomach at 6 o'clock, and the same quantity of same kind of food, masticated in the same manner, into an ounce of simple water; placed them, both together, first in the axilla, and afterwards on the bath, between 96° and 100°.

At 12 o'clock, M., stomach nearly empty. Was just able to get out one ounce for comparison, almost completely dissolved; a few small particles of bread only visible. Temperature 100°.

At 12 o'clock, 30 mins., no distinct particles of food to be seen. All chymified, and passed from the stomach. Nothing but a little frothy mucus remaining in the stomach. Coats clean; colour, pale pink. Temperature 100°.

At 2 o'clock, P. M., he dined on the same quantity and kind of food that he had taken for his breakfast, (*broiled mutton* and *bread.*) Drank nothing since morning. Temperature of stomach 100°. Th: 62°, Wind S. Weather fair, since 12 o'clock. 2 o'clock, 30 mins., stomach as full of fluids as when he drank a pint immediately after eating. No perceptible difference in appearance. 6 o'clock, stomach empty and clean. 9 o'clock, temperature of the stomach 100°. Weather same as at 2 o'clock.

The changes effected in the contents of the two vials, mentioned above, and kept in the axilla till 9 o'clock, P. M., were as follows.

In that containing the gastric juice, the food was about half dissolved, and loosely suspended towards the bottom of a reddish-grey coloured fluid.

That in the water exhibited no other appearance of digestion than what was effected by mastication, when first put in. The masticated food had subsided to the bottom of a transparent, watery fluid, as clear as when first put in.

At 8 o'clock, A. M. of the 28th, I added the two drachms of gastric juice, taken from the stomach, at that time, to the vial containing the gastric juice and the same quantity of water to the watery mixture; and placed them in the axilla again.

At 6 o'clock, P. M., examined vials--digestion had re-commenced, and advanced in the gastric juice, in proportion to the quantity added. The sediment had become more dissolved, and the fluid part increased. This sediment taken out, filtered through muslin, and pressed as dry as when put in, weighed forty-five grains only, having completely dissolved one drachm and fifteen grains; and produced a gruel-like milky coloured fluid.

That in the water, remained unchanged; and when taken out, and pressed dry, through a piece of muslin, like the other, weighed one drachm and thirty-five grains. This reduction, I suppose, was the effects of mastication, and maceration in the water for thirty-six hours.

These two parcels, kept tight corked, in a temperature between 50° and 70°, remained free from any fœtor for forty-five days. The gastric portion, at the end of this time, emitted a caseous flavour; and the aqueous portion smelt musty and sour.

This is a comparison between solution by the gastric juice, and maceration in water. These results are interesting, not only as establishing physiological principles on certain data; but they have an important practical application. They have, consequently, been frequently repeated.

The fact, that the stomach contains a quantity of fluid, soon after the ingestion of dry food, which was alluded to in the preliminary essay, is here perfectly demonstrated.

Experiment 29.

Dec. 28. At 8 o'clock, A. M. Weather clear. Atmosphere dry. Wind N. Th: 34°. Temperature of stomach, 100°. Coats clean and healthy. Gastric juice scarce; extracted two drachms only, and that with considerable difficulty.

At 9 o'clock, A. M., he breakfasted on same kind of food as yesterday, in usual manner, slightly masticated, and swallowed fast, without regard to quantity. 1 o'clock, P. M., a small portion still in the stomach---nearly dissolved. 1 o'clock, 30 mins. stomach empty.

Experiment 30.

Dec. 29. At 9 o'clock, A. M. Weather clear and dry. Wind N. W. and light. Th: 34°. Temperature of stomach, 100°. Coats clean and healthy. He breakfasted on *fat pork, dry toast* and *coffee*--full meal. 1 o'clock, P. M., stomach half full of a lardaceous fluid---no particle of any thing else but gastric fluids to be seen. Temperature 100°. 2 o'clock, 30 mins., stomach not empty. 3 o'clock, stomach empty and clean.

The protracted period of complete chymification in this meal, I conceive to have been principally owing to the unusual *quantity* of food taken, being disproportioned to the gastric secretions, and more than was required to replenish the natural waste of the system. The *quality* of the food had, undoubtedly, some effect.

Experiment 31.

Dec. 30. At 8 o'clock, A. M. Weather clear and dry. Wind N. W. and light. Th: 26°. Stomach clean and healthy. Temperature 100°. Gastric juice pure, and distills more freely than common Extracted one ounce, without any difficulty.

At 9 o'clock, he breakfasted on two and a half ounces of *boiled, recently salted, fat pork*, three ounces of *wheat bread*, masticated in usual manner, and one pint of *coffee.*

At the same time, I took two parcels, equal quantities, of the same kind of food, (pork and bread) half a drachm of each kind, both masticated in same manner: put one of them into the ounce of gastric juice taken from the stomach before eating; and the other, into the same quantity of simple water, of the temperature of the gastric juice; and placed them in the axilla.

At 11 o'clock, I took out of the stomach, one and a half ounces of its contents; put it into a vial, and placed it in the axilla, with the other two. The difference between this taken out of the stomach, and that in the gastric juice, was quite perceptible. The particles of aliment contained in the last, appeared more nearly dissolved, very few remaining distinct. That taken from the stomach contained a larger proportion of the entire food and floating oil. The colour of the middle portions, as well as the smell and taste, were very similar. That from the stomach was rather more rancid and sharp than that in the gastric juice in the vial. Both possessed the peculiar gastric, acid flavour.

At 1 o'clock, 30 mins., the stomach was empty and clean, and probably was so at 1 o'clock; but owing to accident, I did not examine at that time. He became intoxicated in the afternoon, and interrupted the experiments.

On the 2d of *January*, 1833, I added half an ounce of fresh gastric juice to the parcel of chyme taken from the stomach at 11 o'clock, in the above experiment, which, at this time, contained a large proportion of undigested lardaceous matter, floating on the surface. Put the vial in the axilla.

On the 3d, I added three drachms more of fresh gastric juice, to the above.

On the 6th, I added three drachms gastric juice to the above, and placed it on the bath.

On the addition of each of these portions of gastric juice, chymification recommenced, and the lardaceous portion of the aliment continued to be reduced for several hours, till the solvent power became expended, when its action would cease.

Experiment 32.

Dec. 31. At 7 o'clock, A. M. Weather cloudy. Atmosphere damp and chilly. Wind S. Th : 30°. Temperature of the stomach, 100½°—colour darker red than natural, and arid. Mucous coat abraded in spots, and rolled in small shreds; more irritable than usual.

At 8 o'clock, 30 mins., breakfasted on same quantity and kind of food as yesterday, (pork, bread, &c.) At 11 o'clock, took out one and a half ounces contents from the stomach, in appearance half digested. 12 o'clock, M., took out another portion, more completely dissolved. Stomach nearly empty. 1 o'clock, stomach empty.

At 1 o'clock, 30 mins., he dined on *salted, boiled beef, potatoes, parsnips* and *bread*, full meal, without regard to quantity or mastication. 4 o'clock, 30 minutes, stomach perfectly empty.

The one and a half ounces, taken from the stomach at 11 o'clock, A. M., very nearly resembled the contents of the vial of gastric juice and masticated food of the 30th, (yesterday,) in almost every particular. That taken out at 12 o'clock, M., had more of the lardaceous, and less of the distinct fibrous particles of aliment.

The diseased appearance of the stomach at this examination, was probably the effect of intoxication the day before.

Experiment 33.

Jan. 1, 1833. At 8 o'clock, A. M. Weather dark and rainy. Wind S. Th: 50°. Temperature of stomach, 100°, healthy and clean. Extracted half an ounce of gastric juice.

At 9 o'clock, I took two scruples *salted, lean beef,* (boiled,) chopped very fine, with a knife: put one scruple into the half ounce of gastric juice, and the other scruple into half an ounce of simple water; and placed them together in the axilla. At the same time, he breakfasted on two ounces of *boiled, salted, lean beef, bread,* and a pint of *coffee.*

At 12 o'clock, M., I took from the stomach one ounce of its contents, not fully digested; bread principally remaining, reduced to a pulp. Compared with the gastric juice and food in the vial, the particles of meat seemed rather more dissolved. Stomach about half empty.

At 1 o'clock, P. M., stomach empty and clean.

At 8 o'clock, 30 mins., A.M, on the 3d, I added one drachm fresh gastric juice to the vial of gastric juice and chopped beef, and one drachm of water, to the watery mixture, and placed them together in the axilla.

On the 4th, the beef in the gastric juice not being completely dissolved, I added two drachms fresh gastric juice to it; and two drachms of water to the aqueous mixture. Continued them on the bath, or in the axilla. The watery portion began now to smell quite fœtid.

At 8 o'clock, on the 5th, the meat in the gastric juice was completely dissolved, and a fine, reddish grey sediment had fallen to the bottom of an opaque, gruel-like fluid, with a pellicle of greyish white particles on the top. The aqueous portion had become more fœtid. The particles of meat were the same as when first put in, only a little macerated, and paler—the fluid transparent, but becoming darker, and a little greenish—no appearance of solution.

On the 10th, the contents of the aqueous portion were quite fœtid. The gastric portion was perfectly sweet and bland.

Experiment 34.

At 1 o'clock, 30 mins., P. M. same day, he dined on *lean, salted beef* and *bread.* Digested in three and a half hours.

Experiment 35.

Equal parts of *alcohol* and *gastric juice,* mixed together and agitated, produced a turbid, milky white fluid; which, after standing at rest, raised a thin, white coat of fine, loose coagulæ on the surface. When the juice and alcohol were first put together, and before agitating, the gastric juice settled to the bottom, and the alcohol remained on the top, indicating that its specific gravity was less than the fluid.

Experiment 36.

Jan. 2. At 8 o'clock, A. M.—stomach empty—extracted half an ounce of gastric juice. 8 o'clock, 30 mins., he breakfasted on *dry bread* and a pint of *coffee.* 11 o'clock, stomach nearly full of a pulpous, semi-fluid mass, resembling thick gruel. 12 o'clock, nearly empty. 12 o'clock, 30 mins., empty and clean.

Experiment 37.

At 2 o'clock, P. M., he dined on *boiled potatoes,* a small piece of *bread,* and drank a glass of water. 4 o'clock, 30 mins., stomach full of fluids, and quite acrid, of a whitish colour, with particles of potatoes floating about. 6 o'clock, stomach empty.

Experiment 38.

Jan. 3. At 8 o'clock, 30 mins., A. M. Weather pleasant, smoky and clear. Th: 38°. Temperature of the stomach, 101½°, immediately after a walk of two miles, producing free perspiration, and colour in the face. Extracted half an ounce of gastric juice.

At 9 o'clock, 30 mins., he breakfasted on cold *broiled breast of veal, boiled potatoes*, and *bread*. At the same, or within fifteen minutes of the time, I suspended into the stomach, at the aperture, twenty grains of masticated lean veal, contained in a muslin bag.

At 12 o'clock, M., contents of stomach half diminished. 1 o'clock, P. M., stomach nearly empty. 1 o'clock, 30 mins., all gone from the stomach, except the muslin bag and contents. The contents appeared to be about half diminished.

At 2 o'clock, I took out the bag of veal, and pressing it as dry as I could, without forcing the remaining particles of meat through the cloth, it weighed ten gains, having lost ten grains by digestion, in four and a half hours. The veal, when first put in the bag, and suspended in the stomach, was of a clay, or greyish white colour; but when taken out and weighed, was of a palish red, or light flesh colour, and of a glutinous appearance.

Experiment 39.

At 3 o'clock, P. M., same day, dined on *broiled veal* and *bread*, and drank half a pint of *water*. Digested in two hours.

Experiment 40.

Jan. 4. At 8 o'clock, A. M.—Stomach healthy. Extracted two drachms gastric juice—came pure, but very slow.

At 9 o'clock, breakfasted on *broiled veal*, *bread* and *coffee*. 11 o'clock, stomach full—oil floating on the surface, acrid and sharp, excoriating the edges of the aperture and skin. 12 o'clock, M., chyme passing out. Stomach two thirds empty. 1 o'clock, P. M., stomach empty.

Experiment 41.

At 2 o'clock, P. M., same day, he dined on *breast of broiled veal* and *bread*, and drank a tumbler of *water*.

5 o'clock, 30 mins., stomach nearly empty. 6 o'clock, examined stomach—chyme of a milky white colour. 6 o'clock, 30 mins., chyme still remaining. 7 o'clock, stomach not empty. Took out half an ounce of contents. It was a milky white fluid, with a peculiar smell, and slightly acid and bitter taste. 7 o'clock, 15 mins., stomach empty.

Experiment 42.

Jan. 5. At 8 o'clock, A. M.—Stomach healthy and clean. Extracted half an ounce of gastric juice. Put it into a vial, and immersed in it fifteen grains of firm *tendon* of young beef, in a solid piece. Kept it either in the axilla, or on the bath, for twenty-four hours, when all was completely dissolved.

At 8 o'clock, 45 mins., he breakfasted on *broiled veal, bread* and *coffee*, and kept exercising. 12 o'clock, M., stomach about half empty. Took out half an ounce, completely dissolved—no distinct particles of food to be seen. 12 o'clock, 30 mins., M., all gone.

This affords an example of the digestion of tendon. Hard, solid substances require a greater quantity of gastric juice than more tender fibre, and take a longer time for their complete solution.

Experiment 43.

At 1 o'clock, P. M., same day, dined on *broiled veal* and *bread*, and drank half a pint of *water*. Digestion completed in four and a half hours.

Experiment 44.

Jan. 6. At 8 o'clock, A. M.—Examined stomach. Coats generally healthy—few small, erythematous patches, on mucous surface. Secretions pure. Extracted one and a half ounces clear gastric juice, containing less than the usual quantity of mucous flocculi. It ran more freely than common through the tube. More could have been obtained; but a sensation of faintness, and sinking at the pit of the stomach, being felt and complained of, I desisted. This sensation has almost uniformly occurred, whenever the gastric juice has flowed more freely than usual, and has been suffered to run out to the quantity of one and a half, or two ounces; followed by dimness of vision, and vertigo, on rising. These feelings, however, subside in a few minutes, and he feels as usual, and eats his meals with a good appetite.

At 9 o'clock, he breakfasted on *broiled veal* and *bread* again, as yesterday, and kept exercising. 1 o'clock P, M., stomach nearly empty—several small spots of dark, grumous blood, exuding from the papillæ of the inner coats, made their appearance. 2 o'clock, some appearance of the breakfast still in the stomach. 2 o'clock, 15 mins., stomach empty.

Experiment 45.

At 2 o'clock, 30 mins., P. M., same day, he dined on one pint of *barley gruel*, sweetened with molasses. 4 o'clock, 30 mins., stomach empty—none of the barley gruel to be seen.

Several small, sharp pointed, white pustules made their appearance on the inner surface of the stomach, at this time; and the surface, generally, was of a paler colour, and more flaccid, than usual.

Experiment 46.

Jan. 7. At 8 o'clock, A. M.—Weather cloudy, damp, and disagreeable. Th: 48°. Wind N. E. Temperature of stomach, 100°. Less of the small pustules and red patches than yesterday. Colour of the coats natural again; but little secretion of gastric juice this morning. Could obtain only a drachm or two.

At 9 o'clock, A. M.—Temperature of stomach, 100°. He breakfasted on *soft boiled eggs*, *soft toast*, and *coffee*. 12 o'clock, M., stomach empty.

Experiment 47.

At 12 o'clock, 30 mins., M., same day, he dined on three *hard boiled eggs* and *bread*. 3 o'clock, 30 mins., stomach half empty. Remaining contents acrid. Edges of the aperture excoriated. Some pimples, and erythematous patches on the surface of the inner coats. 4 o'clock, 30 mins., stomach and contents in nearly the same condition as at last examination—very acrid and sharp—coats red. 6 o'-clock, stomach empty.

These three or four last experiments demonstrate, that a diseased state of the stomach retards digestion.

Experiment 48.

Jan. 8. At 8 o'clock, 30 mins., A. M.—Examined stomach. Coats healthy. None of those white pustules, and erythematous patches, observed yesterday and the day before, to be seen this morning. Colour of the lining membrane rather paler than common. Surface moist. Extracted half an ounce of gastric juice, without difficulty. A slight and momentary vertigo was felt in rising up. No faintness or sense of sinking at the scrobiculus cordis, at this extraction. I divided these four drachms of gastric juice into two, equal parts, and put them into separate vials. In a third vial, I put two drachms of simple water. To each of these three vials, I added eleven grains of the muscle of a *sheep's heart*, in an entire piece. Kept one of the vials of gastric juice and meat in the axilla, and placed the other, with the aqueous vial, in a cool place, at about 46°, agitating them alike frequently.

At 7 o'clock, P. M., the piece in the warm gastric juice was half digested; the fluid of an opaque, reddish brown colour. That in the cold gastric juice was a very little affected, the surface being covered with a thin, glutinous coat, and the fluid a little turbid. That in the water was not in the least affected.

The water was perfectly transparent, as when first put in.

At 9 o'clock, A. M., of the 9th, these several pieces of muscle exhibited the following results. That in the warm gastric juice, when taken out and pressed dry, as when put in, weighed seven and a half grains. That in the cold gastric juice, treated in the same manner, weighed twelve and a half grains, having *gained*, by the absorption of gastric juice, one and a half grains. And that in the simple water, weighed eleven grains, the same as when put in, having neither lost nor gained.

The three and a half grains, that remained in the first vial, were in one entire piece, of the same shape as when first put in; but very soft and tender, hardly able to sustain sufficient pressure to be raised by the finger and thumb. It was a mere pulp.

The meat in the second vial was increased a little in size; appeared swollen, soft, slimy and tender; but had sufficient firmness of texture to resist considerable pressure, when taken up. It was not dissolved.

That in the water retained its firmness, and was unaltered in appearance, except a paleness of surface, occasioned by maceration.

At 8 o'clock, next morning, (the 10th,) the following appearances were evident.

The first piece, in the warm gastric juice, weighed one and a half grains, having lost in the last twenty-three hours, two grains only. It retained the same shape, and was of about the same consistence as yesterday. A reddish brown sediment subsided to the bottom of a rich, whey-coloured fluid.

The second piece, in the cold gastric juice, weighed nine grains and a fraction, having lost about three and a half grains.

That in the water, was unaltered, and weighed the same as when put in—eleven grains.

It may be proper to remark, that the two pieces in the cold gastric juice and water, were moved from their first position in a temperature of about 46°, and placed for the last twenty-three hours on the mantle-piece, over the fire, in my room, in a temperature of about 60°.

The loss of the two and a half grains of meat, in the cold gastric juice, was evidently the effect of digestion, occasioned, no doubt, by the increase of fourteen or fifteen degrees of temperature.

On the 10th, I added to the vial, containing the warm gastric juice and muscle, one fourth of a drachm of fresh gastric juice, warm from the stomach. Continued it in axilla, and in five hours it was dissolved to a mite, scarcely perceptible.

The piece in the cold gastric juice, kept on the mantle-piece, in a temperature between 50° and 60°, till 9 o'clock, A. M., of the 11th, weighed seven grains, retaining the same shape as yesterday, and a similar texture. The fluid had become more opaque and milky, and the sediment had increased at the bottom.

The piece in the water at this time, remained unaltered, and weighed precisely the same as at first—eleven grains.

At 9 o'clock, A. M., I placed both these in the axilla.

At 9 o'clock, P. M., the piece remaining in the second vial of gastric juice, placed in the axilla this morning, was nearly all dissolved, one grain only remaining—a soft pulp.

The piece in the water remained unaltered, and weighed the same as at first; but began to emit a strong fœtid odour, and in a few days became very putrid. This was, however, almost entirely corrected, by the addition of three drachms of fresh gastric juice on the 21st. The meat still continued its original shape and size, and no doubt, its weight, though too putrid to handle, or take out, before the addition of the gastric juice. Placed it on the bath, and it began to digest, and soon became chymified—lost its fœtid smell, and acquired a sharp acid, or rather, acrid taste.

The result of this experiment is interesting, in demonstrating the solvent properties of the gastric juice. Maceration alone will not dissolve food, nor separate its nutritious parts. It appears, also, from this experiment, that gastric juice corrects the putrid tendency of aliments; and that food is more readily dissolved after that tendency has occurred.

Experiment 49.

Jan. 11. At 8 o'clock, A. M.—Weather clear and dry. Wind S. W. Th: 15°. Temperature of the stomach, 100°. Coats healthy. Extracted one ounce of gastric juice, clear and transparent—few flocculi of mucus—taste distinctly acid. Complains of the usual sense of distress at the pit of the stomach, and vertigo.

At 9 o'clock, 30 mins., he breakfasted on *pork* and *bread.* Digested in four hours and a half.

Experiment 50.

At 9 o'clock, 30 mins. A. M., same day, I took three vials, and put into each two drachms pure gastric juice, fresh from the healthy stomach. To one, I added one drachm of *albumen*—white of egg—to the second, half a drachm of the *yolk*—and to the third, another drachm of *albumen.* Put the two first, in axilla, and the other on the mantle-piece.

At 9 o'clock, P. M., the albumen in the warm gastric juice, in the axilla, had become quite opaque, with loose, light coloured sediment at the bottom. The albumen in the cold gastric juice remained unaltered. That containing the yolk, exhibited the appearance of a mere mixture of fine yellow coagulæ, resembling sulphur and milk, mixed together.

On the 12th, at 8 o'clock, P. M., both vials having been continued on the bath, or in the axilla, through the day, the difference observed last evening, between the cold and warm vials of albumen, was very little increased.

The yolk was considerably altered from a loose coagulæ, generally diffused through the gastric juice, to a fine compact body of coagulæ, rising upon the top of a perfectly clear, transparent fluid, free from a particle of sediment.

Experiment 51.

At 8 o'clock, 30 mins., A. M.—Stomach healthy. Extracted one ounce of gastric juice, a little tinged with yellow, whether from bile or tobacco, it was difficult to determine. He had taken some tobacco into his mouth, an hour and a half previous to the examination, and the fluid was not perceptibly bitter. There was a larger portion of frothy saliva, and flocculi of mucus, than common.

At 10 o'clock, 15 mins., he breakfasted on *boiled, salted codfish, bread* and *coffee.* Digested in two hours and a quarter.

Experiment 52.

Jan. 13. At 8 o'clock, A. M.—Weather overcast, dry and smoky. Light wind. Th : 12°. Temperature of stomach, 100° and a fraction. Pulse 60, in a recumbent, and 70, in an erect position. Coats not perfectly healthy—general surface rather paler than usual—some red spots and pimples to be seen. Extracted three drachms of gastric juice, slightly acid—not so much as usual—less mucus, and more saliva than common. Neither tinge nor taste of bile.

At 9 o'clock, he breakfasted on *boiled, fat pork* and *bread.*

At 12 o'clock, M.—Stomach two thirds empty.—Temperature, 100° and a fraction.

At 2 o'clock, P. M.—Stomach nearly empty—very little pulp of bread, and lardaceous fluid to be seen. Has just returned from walking two miles or more. Temperature of stomach, 100½°.

At 1 o'clock, 30 mins., stomach empty. Temperature, 101°.

Experiment 53.

Jan. 9. At 2 o'clock, P. M. same day, he dined on *boiled, fat pork, boiled cabbage* and *bread*, and drank a tumbler of *water*. Digested in five hours. 9 o'clock, temperature 100°.

Experiment 54.

Jan. 14. At 8 o'clock, 40 mins., A. M.—Weather clear, dry and serene. Wind N. W. and light. Th: 28°. Stomach healthy. Coats clean. Temperature of stomach, 100°. Extracted nine drachms of pure gastric juice--distinctly acid--few flocculi of mucus, and a little appearance of frothy saliva. A slight sense of faintness and vertigo ensued, as usual, on rising, after this quantity.

At 9 o'clock, breakfasted on *boiled, fat pork* and *bread.* 12 o'clock, M., stomach about half full. Temperature, immediately after walking two and a half miles, 101½. 1 o'clock, P. M., stomach empty and clean. Temperature 100°.

Experiment 55.

At 2 o'clock, P. M., same day, he dined on *boiled, fat pork* and *bread.* Digested in three hours.

Experiment 56.

Jan. 14. At 9 o'clock, A. M., I put a solid piece of *rib bone*, of an old hog, weighing ten grains, into a vial, containing three drachms of pure gastric juice, taken from the stomach this morning. Placed it in the axilla, and continued it there for twelve hours; then placed it on the shelf, in a cool place, till next morning.

-- 15. 9 o'clock, A. M., surface of bone evidently dissolved. Fluid quite opaque. Took out the piece; and when wiped, and dried with blotting paper, as dry as when put in, it weighed just nine grains.

Immersed it again in the same juice, and placed it on the sand bath at 100°. Continued it in that temperature for twelve hours, frequently agitating it; then, as yesterday, placed it on the shelf, until next morning.

— 16. 9 o'clock, A. M., appearance similar to yesterday morning. Juice a little more turbid. Bone covered with a thin, cineritious coat. Taken out and wiped, the piece weighed eight and a half grains. Immersed again in same fluid, and continued on bath twelve hours; then set on shelf again until next morning.

— 17. 9 o'clock, A. M., very little alteration since yesterday. Bone taken out and wiped, weighed eight and a half grains. Put in again, and continued on bath fifteen hours.

— 18. 12 o'clock, M., no change effected since last examination. Bone taken out and wiped, weighed precisely same as yesterday, eight and a quarter grains. Conceiving the solution of the bone had ceased from a deficiency of the gastric solvent, I now added one drachm fresh gastric juice, and continued it on the bath again, for eight hours.

— 19. 12 o'clock, M., bone taken out, and wiped, as usual, weighed eight grains. Returned to bath, and continued twelve hours, it weighed seven and a half grains. Returned, and continued on bath thirty-six hours, and frequently agitated, between

—20th and 25th, no visible change was effected. Weight same as on the 19th, seven and a half grains. The solution having ceased again, I added three drachms more of gastric juice, and continued it on the bath twenty-four hours.

— 27. 10 o'clock, A. M., laminæ of bone separated, and opening on one edge. Fluid more opaque, with a little fine, brown sediment, precipitated to the bottom of the vial. Weight of bone, five and a half grains. Added two drachms of gastric juice, and continued it on the bath for eighteen hours.

— 28. 10 o'clock, A. M., laminæ of bone opened. Weight, four grains. Returned, and continued on bath twelve hours.

— 29. 10 o'clock, A. M., laminæ of bone entirely separated, thin as paper, and elastic as horn.— Weight, three and a quarter grains. Returned to bath twelve hours.

— 30. 10 o'clock, A. M., opacity of fluid, and fine sediment, increased. Weight of bone, two and three fourths grains. Continued on bath.

— 31. 10 o'clock, A. M., no change since yesterday. Weight of bone, two and three fourths grains. Added half a drachm of gastric juice, and continued it on bath twelve hours.

Feb. 1. 10 o'clock, A. M., laminæ very thin and elastic. Weight of bone two and a half grains.— Took out the pieces of bone, and put them into one drachm fresh gastric juice, in a separate vial, and continued on bath six hours.

— 2. 10 o'clock, A. M., weight of bone, two and a quarter grains. Continued on bath six hours.

— 3. 10 o'clock, A. M., weight of bone, two grains. Continued on bath till the 5th.

— 5. 10 o'clock, A. M., no change since the 3d. Weight of bone same. Added two drachms gastric juice, and continued on bath twelve hours.

— 6. 10 o'clock, A. M., bones nearly all dissolved—three fourths of a grain only remaining.

— 7. Weight of bone, half a grain, very thin and transparent. The solution not being quite completed, I added two drachms more of gastric juice, and continued it on bath twelve hours.

— 8. 10 o'clock, A. M., all dissolved to a mite— quarter of a grain, or less.

After the solution of the bone, the menstruum was a greyish white, opaque fluid, nearly of the colour and consistence of clear, thin gruel, with considerable fine brown sediment at the bottom of the vial, after standing at rest awhile; and had a peculiarly insipid, sweetish taste, and smell—not the least fœtor or rancidity.

It will be seen, in this experiment, that the piece of bone was dissolved in proportion to the quantity of gastric juice applied, and that the solution ceased at longer or shorter intervals, as a larger or smaller quantity was added. When the juice became saturated, as well as when the vial was removed from the bath to a low temperature, the solution ceased. It appears that it took fourteen and a half drachms of gastric juice to dissolve ten grains of solid bone.

Experiment 57.

Jan. 15. At 8 o'clock, A. M.—Weather cloudy and dry. Wind N. E. and light. Th: 35°. Temperature of the stomach, 100°. At 9 o'clock, A. M., he breakfasted on *fat pork* and *bread.* 2 o'clock, P. M., stomach empty and clean. Temperature, 101°.

Experiment 58.

At 2 o'clock, P. M., same day, I put fifteen grains of *raw beef steak*, divided into small pieces, into three drachms of gastric juice; and fifteen grains of *broiled beef steak*, into other three drachms of gastric juice. At the same time, I put the same quantity of *broiled steak*, divided like the others, into three drachms of saliva, fresh from the mouth. I then placed them, all together, alternately in the axilla and on the bath, and kept frequently agitating them.

At 4 o'clock, the meat in the saliva exhibited the appearance of simple maceration; the other two parcels, in the gastric juice, were considerably diminished and partially dissolved, the fluid of an opaque, whitish colour; the cooked piece, rather the most dissolved.

At 6 o'clock, the salivary portion was not much changed in appearance; the other two about half dissolved; the cooked meat in advance of the raw.

At 9 o'clock, the salivary portion began to smell slightly fœtid, and to change colour. The other two were perfectly bland, and of a sweetish flavour—the meat about three fourths dissolved, with a fine, brownish red sediment at the bottoms of the vials. Took them all off the bath, and placed them on the shelf till next morning.

At 7 o'clock, A. M., on the 16th, I placed them again on the bath till 9 o'clock, when the salivary portion had become fœtid, and was of a greenish colour. The fibres of the meat retained their shape and size; and had become pale on the surface. Light, loose coagulæ had fallen to the bottom, leaving a reddish green coloured fluid above. The gastric portions were almost completely dissolved; the cooked meat still in advance.

At 12 o'clock, M., the salivary portion was very fœtid. The remaining portions of aliment, taken from the three vials, filtered through thin muslin, and dried with blotting paper, weighed as follows:—the broiled meat, in gastric juice, one grain; the raw meat, in the same, two and a half grains; and that in the saliva, twelve grains.

This experiment demonstrates that saliva does not possess the properties of a solvent; but facilitates putrefaction. See, also, subsequent experiments. It also shows, that raw meat is susceptible of digestion by the gastric juice, though in a less degree than cooked meat.

Experiment 59.

Jan. 17. At 9 o'clock, A. M.—Weather clear, and dry. Wind N. W. and light. Th: 19°. Temperature of stomach, 100°. Coats clean and healthy. Extracted ten drachms of gastric fluid; not so clear and limpid as usual; some streaks of yellow bile, and more appearance of saliva than common—acid not so perceptible as usual. I divided this into three equal parts, three and one third drachms each. To one part, I put fifteen grains firmly coagulated *albumen*, (white of egg)—to the other, fifteen grains of the soft coagulæ of the same—and to the third, fifteen grains raw *albumen*—and placed them on the bath and in axilla, alternately.

At the same time, he breakfasted on three *hard boiled eggs*, *bread* and *coffee*.

At 11 o'clock, examined—stomach full. Temperature, 100°. Some small red spots. Contents acrid.

At 12 o'clock, M.—just returned from walking one mile, and back again. Weather clear, dry and serene. Wind N. W. and light. Th: 23°. Temperature of stomach, 102°—nearly empty. Took out one ounce, almost completely chymified; a little pure oil floating on the surface. Put this on the bath.

At 12 o'clock, 30 mins., stomach empty.

At 9 o'clock, P. M., examined the parcels of albumen, placed in the vials of gastric juice this morning, at 9 o'clock. Of the firm coagulæ, there remained one and a quarter grains; of the soft, none; of the raw, three fourths of a grain, in loose, white coagulæ.

Experiment 60.

Jan. 17. At 12 o'clock, 30 mins., M. I put twenty-five grains *lean, broiled mutton*, divided into small pieces, into five drachms of gastric juice, and same quantity into five drachms of gastric juice and fresh saliva, mixed together; and placed them on the bath.

At 9 o'clock, P. M., the meat remaining in the gastric juice, taken out and dried with paper, weighed just twelve grains; that in the mixture of gastric juice and saliva, weighed eighteen and three fourths grains. The texture of the first was considerably more dissolved and tender than the second. Returned them into their respective vials again.

At 12 o'clock, 30 mins., M. of the 18th, examined them again. The meat remaining in the gastric juice, weighed five and three fourths grains; was soft, glutinous, and of a dirty brown colour. That in the gastric juice and saliva, weighed thirteen and a quarter grains; the texture was quite firm, and retained its fibrous form, and reddish, bloody colour. Put them in the bath again.

At 4 o'clock, P. M., of the 19th, the meat in the gastric juice weighed two grains. Consistence and colour of fluids, same as yesterday. The meat in the gastric juice and saliva, weighed nine and a half grains. Fluids of a reddish brown colour, and less precipitate.

In ten days, the salivary mixture became very putrid; but the gastric portion was perfectly sweet, and so continued for thirty days, or more.

Experiment 61.

Jan. 18. With a view to ascertain the antiseptic properties of the gastric juice, I took a portion of very putrid animal matter, and added to it a quantity of gastric juice. The fœtor was at once almost completely corrected, leaving only a slight putrescent smell, with the usual flavour of the gastric juice.

Experiment 62.

At 9 o'clock, A. M., same day, extracted one and a half drachms of gastric juice, and added it to two and a half drachms of *milk*. The whole was formed into loose, white coagulæ, in less than five minutes. At 1 o'clock, P. M., remaining coagulæ, after filtering through muslin, weighed thirteen grains. Returned it into the vial, and placed it on the bath again. At 9 o'clock, no coagulæ remaining—all completely dissolved.

Experiment 63.

Jan. 9. At 9 o'clock, A. M., coats of stomach perfectly healthy and clean. No appearance of morbid action—tongue clean—and every indication of perfect health. There was no free fluid in the gastric cavity, until after the elastic tube was introduced, when it began slowly to distill from the end of the tube, drop by drop, perfectly transparent, and distinctly acid. I obtained about one drachm of this kind, and then gave him a mouthful of bread to eat. No sooner had he swallowed it, than the fluid commenced flowing more freely from the tube, and I obtained two drachms, less pure, however, with saliva and mucus mixed with it, and slightly tinged with yellow bile. The surface of the protruded portion of the villous coat at this time became covered with a limpid fluid, uniformly spread over its whole surface, distilling from myriads of very fine papillary points, and trickling down the sides. After letting him rise and walk about two or three minutes, I again introduced the tube, and obtained about two drachms more of very pure gastric juice, making, in the whole, five drachms.

Breakfasted on *boiled pork* and *bread.* Dined and supped on the same.

Experiment 64.

Jan. 20. At 8 o'clock, 30 mins., A. M., examined stomach—appearances healthy. Extracted three drachms gastric fluids, coloured with bile, slightly acid and bitter. It ran more freely than yesterday.

At 8 o'clock, 45 mins., he swallowed four ounces of pure *gelatine*, (ichthyocolla,) prepared with boiling water, transparent, and of a tremulous consistence.

At 9 o'clock, stomach appeared nearly as full as usual after eating his ordinary meals ; fluid, clear, and of the consistence of the albumen of eggs. It appeared to be the gelatine, dissolved, or diffused in the gastric juice. The juice and the liquid gelatine so much resembled each other, that I could not distinguish them apart.

At 9 o'clock, 45 mins., examined again—found the stomach almost entirely empty—was just able to obtain two drachms of fluid. It appeared to be a mixture of gelatinous chyme, gastric juice and flocculi of mucus, more opaque and ropy than the gastric juice alone, and more acid than the fluids of the stomach immediately before the gelatine was swallowed. Not the least appearance of bile or yellow colour in the gastric cavity, or fluids, after taking the gelatine: considerable vertigo followed the extraction of this last fluid. It soon passed over, however, and he ate his breakfast, (*pork* and *bread*) with his usual appetite.

The process of the solution of gelatine, is difficult to ascertain. It is not subject to coagulation; and the action of the gastric juice is not easily perceived. It is no doubt dissolved by the gastric juice, in the same manner as other aliment is. See subsequent experiments.

Experiment 65.

To ascertain whether the sense of hunger would be allayed without the food being passed through the œsophagus, he fasted from breakfast time, till 4 o'clock, P. M., and became quite hungry. I then put in at the aperture, three and a half drachms of *lean, boiled beef*. The sense of hunger immediately subsided, and stopped the borborygmus, or croaking noise, caused by the motion of air in the stomach and intestines, peculiar to him since the wound, and almost always observed when the stomach is empty.

This experiment proves that the sense of hunger resides in the stomach, and is as well allayed by putting the food directly into the stomach, as when the previous steps have been gone through with. Not that I would deny the utility of the previous processes, in ordinary cases. Even the sense of taste is essential. It is placed as a sentinel, to prevent improper articles from being introduced into the stomach. See, also, subsequent experiments.

Experiment 66.

Jan. 21. At 8 o'clock, A. M.--Examined stomach. Could obtain but few drops of gastric juice. Sent him to exercise in the open air for half an hour. Secretions increased--gastric juice flows pure, and more freely. Extracted three drachms.

At 8 o'clock, 30 mins., he breakfasted on *bread* and *coffee*, and a small piece of *lean pork.*

At 2 o'clock, P. M., stomach empty. Extracted two drachms of gastric juice, tinged with yellow bile; and then one drachm of pure, transparent juice, distilling, by drops, from the end of the tube.

At 2 o'clock, 30 mins., I put ten grains of *raw suet* into two drachms of gastric juice, tinged with bile, and ten grains of the same, into two drachms of pure gastric juice. Placed them both on the bath.

At 9 o'clock, the piece of suet in the juice that was tinged with bile, was considerably more dissolved than that in the clear gastric juice; and when examined with the compound microscope, the globules appeared more numerous and much smaller. This appearance was, also, clearly perceptible to the naked eye, as the mixtures stood in the vials.

At 10 o'clock, the piece in the yellow juice was all dissolved—the other, not entirely.

This, with other subsequent experiments, indicate that oily or fatty food is sooner digested when there is a small admixture of bile with the gastric juice.

Exercise, it seems, promotes the discharge of the gastric juice, as well as digestion in the stomach.

Experiment 67.

Jan. 22. At 8 o'clock, 30 mins., A. M.--Stomach clean and healthy. Extracted five drachms of very clear, pure gastric juice. The first three drachms ran out quite freely; the other two drachms distilled by drops. It was not the least tinged with bile, and tasted distinctly acid. Breakfasted on *beef steak*, *bread* and *coffee*.

At 1 o'clock, P. M., stomach empty.

Experiment 68.

At 9 o'clock, P. M., same day, St. Martin having eaten nothing since 2 o'clock, and feeling quite hungry, I put into the stomach, at the aperture, eight ounces of beef and barley soup, introduced gently through a tube, with a syringe, lukewarm. It caused no unpleasant sensation, but allayed the sense of hunger. It satisfied the *appetite;* and he said he had no desire to *eat*.

At 10 o'clock, he said he felt a little hungry again, and ate eight ounces more of the same kind of soup, which had a similar effect as the other.

Experiment 69.

Jan. 23. At 9 o'clock, A. M.—Weather rainy.—Wind N. E. and light. Th: 39°. Stomach empty, clean and healthy. Temperature of stomach, 100½°.* Breakfasted on *sausage*, *bread* and *coffee*.

At 10 o'clock—Aspect of weather same as at 9 o'clock. Th: 40°. Stomach full of fluids—temperature 101¾°. The spirit became stationary at that point, after keeping the tube in the aperture eight or ten minutes: after which, it did not vary for ten minutes, when it was taken out.

At 12 o'clock, M., he returned from a walk of two miles. Stomach nearly empty. Temperature 101½°—stationary after being continued five minutes in the stomach.

At 12 o'clock, 30 mins., stomach empty.

* In this, and the subsequent experiments, I used a spirit Thermometer, taken from Pool's Barometer, which varied half a degree from those formerly used.

Experiment 70.

Jan. 24. At 8 o'clock, A. M., Weather cloudy and damp. Wind N. and moderate. Th: 39°. Stomach empty, clean and healthy--temperature, 100½°. Extracted four drachms gastric juice, very little tinged with yellow.

At 9 o'clock, he returned from a short walk. Temperature of stomach, the same. Breakfasted on *bread* and *coffee.* 12 o'clock, stomach empty.

Temperature of the stomach, after walking two miles or more, 101¼°.

Experiment 71.

At 1 o'clock, P. M., same day, St. Martin complaining of being quite hungry, I put into the stomach at the aperture, twelve *raw oysters*, more than middling size. The sensation was allayed, and the appetite satisfied, the same as if swallowed. He was not hungry again till half after 4 o'clock, when he ate a dozen more of the same kind of oysters, with bread.

At 10 o'clock, P. M., stomach empty and clean. Weather damp and rainy. Wind N. E. and brisk. Temperature of the stomach, 99½°. He had been covered in bed, and sleeping, for two and a half hours, from which I awoke him to introduce the Thermometer. He fell asleep again during the examination---only awoke while putting in and taking out the glass tube.

Experiment 72.

Jan. 25. At 6 o'clock, A. M.--Wind Southerly, and light. Th: 36°. Examined stomach before rising from his bed. Temperature 99°. Extracted fifteen drachms gastric fluid. It flowed out unusually free ; was rather more opaque, and contained less flocculi of mucus than common for the quantity. Particles of the bread eaten with his oysters at 4 o'-clock, 30 mins., yesterday, were distinctly to be seen in this parcel of the juice.

At 8 o'clock, 30 mins.—Temperature of the stomach, 100½°. Coats clean and healthy. Th: 38°.

At 9 o'clock, he breakfasted on *raw oysters* and *bread.* 11 o'clock, temperature of stomach, 101°. 12 o'clock, M., he returned from a walk of two miles. Stomach empty. Temperature, 102°.

Experiment 73.

Jan, 26. At 8 o'clock, A. M.—Weather clear and cold. Wind N. W. and light. Th: 30°. Stomach healthy, empty and clean. Temperature, 100½°. Extracted one drachm gastric juice, containing more than usual flocculi of mucus.

At 9 o'clock, he breakfasted on *sausage*, *bread* and *coffee.* 10 o'clock, Th: 34. Temperature of the stomach, 100 3-4°, and full of a heterogeneous fluid. 12 o'clock, M., returned from a walk. Stomach empty —temperature, 101° and a fraction. Weather clear and pleasant. Th: 39°. Wind N. W. and moderate.

From this, and other experiments, it may be clearly inferred, that in the most natural and healthy states of the stomach, there are little or no fluids, of any kind, in the gastric cavity, until excited by aliment or other irritants; and that digestion, under this condition, is the most rapidly and perfectly performed.

Experiment 74.

At 2 o'clock, P. M., same day, he dined on *raw oysters* and *bread.* 5 o'clock, stomach empty.

At 6 o'clock, 40 mins., immediately after drinking a tumbler of water of the temperature of 55°, introduced Thermometer—spirit rose very slowly, and did not become stationary at the natural temperature until the tube had stood in the stomach for thirty-five minutes. 12 o'clock at night, temperature 99 1-2°, after sleeping in bed three hours.

Experiment 75.

Jan. 27. At 6 o'clock, A. M. – before rising from his bed—Weather cloudy and dry—calm--Th: 32°—Stomach empty, clean and healthy—Temperature 99 1-2°, spirits stationary in ten minutes – he swallowed a gill of water at the temperature of 55°, which immediately diffused itself over the interior of the stomach, and discharged some at the aperture, by the side of the stem of the Thermometer, which had not been withdrawn. The spirit immediately fell to 70°; stood at that point one and a half or two minutes; and then began again very slowly to rise. Thirty minutes elapsed after taking the water, before the spirit regained the 99th degree. Before the end of that time, there was no appearance of water in the gastric cavity.

At 9 o'clock, 30 mins., he ate a full breakfast of *fresh, broiled beef*, mostly fat, *bread* and *coffee*, and continued unusually smart exercise, walking for two hours, till he became fatigued, and perspired freely.

At 11 o'clock, 30 mins.—Weather clear. Th : 43°. Just returned from walking. Stomach contained considerable chyme and oil. Aliment about two thirds gone. Temperature, 101°.

At 12 o'clock, 20 mins., M.—Stomach nearly empty; a small portion of the fluid remaining, reduced to a more perfect chymous condition, with less oil, and that in much finer globules. Appeared tinged with yellow, and tasted bitter.

At 1 o'clock, P. M., chyme gone. Very little oil remaining.

At 2 o'clock—Weather unchanged. Temperature of stomach, 101¼°. No chyme to be seen. A few particles of oil still remaining, floating on the surface of a small quantity of fluid, exhibiting considerable spumous froth and mucus.

A circumstance occurred here, not before observed in my experiments, which it may not be unimportant to mention, *i. e.*--the variations of the the temperature observed in moving the Thermometer up or down in the stomach. The spirits in the tube varied proportionably to the length of the stem introduced. When the bulb sank down to the pyloric portion of the stomach, to the depth of six or eight inches, the spirit rose to 101 1-4° : when only immersed two or three inches, it would stand at 100 1-2°, making a difference of three fourths of a degree. These variations were uniformly observed at every thermometrical examination.

Perhaps the difference of indication of the thermometer, may result from a more complete envelopement of the stem in the gastric cavity, at the pyloric examination, and a less one at the splenic. I give the reader possession of the fact, without pretending to account for it, with certainty.

Experiment 76.

At 2 o'clock, 30 mins., P M., same day, he dined on *raw oysters* and *bread.* 4 o'clock, 30 mins., stomach not empty. Food about half gone. Small pieces of heart of oyster, and pulp of bread, to be seen, floating in a thin, pultaceous fluid, quite acrid and sharp--no bitter taste, or yellow colour. Temperature 101 1-2°.

A striking peculiarity in the movement of the spirit in the Thermometer was observed in this experiment. It rose from about 68° to its stationary point, 100 1-2°, in less than five minutes after the bulb was put into the stomach. At last examination, 2 o'clock, it was fifteen minutes in making the same range. Sometimes it has been twenty-five or thirty minutes before it became stationary, and under no appreciable difference of circumstances. He had been moderately exercising, (walking) immediately before the last examination.

At 5 o'clock, he returned from walking. Temperature of stomach, 101 1-2°. Spirit rose, and became stationary, at that point, in less than three minutes—food almost completely chymified, and half gone. Took out one ounce of thick, pultaceous, porridge-like fluid, with some small pieces of the hearts of the oysters, reduced to a jelly-like appearance—plainly acid—and slightly bitter; and had the flavour of the oysters.

At 6 o'clock, 15 mins., stomach empty and clean.

At 6 o'clock, 30 mins., he ate a full meal of *cold, boiled beef* (considerable fat) and *bread*. 10 o'clock, 30 mins., stomach empty.

Experiment 77.

At 9 o'clock, A. M., of the 27th, I mixed one drachm of the clear *decoction of coffee* with three drachms of fresh gastric juice, with a view to ascertain whether it would destroy the flavour of the coffee. It had no perceptible effect. The flavour of coffee remained for ten hours, as distinct as at first. Added half a drachm of loaf sugar to the mixture, and placed it on the bath. It remained there forty-eight hours: no different effect was produced on the flavour of the coffee. It remained the same as at first.

It is probable that the decoction of coffee, like many other artificial drinks, does not admit of digestion; possesses no nutritive principles; and is carried into the circulatory system without much change.

Experiment 78.

At 1 o'clock, 30 mins. P. M., of the 27th, I put fifteen grains firm beef *cartilage* into three drachms of gastric juice, and placed on bath.

At 10 o'clock, A. M., of the 28th, took out and wiped dry, it weighed six and three fourths grains.

At 10 o'clock, A. M., of the 29th, it weighed one grain.

When put in, the cartilage was cut into different sized pieces: these retained their original forms till completely dissolved—the largest piece being the last digested.

Experiment 79.

Jan. 28. At 6 o'clock, 30 mins., A. M.—before rising.—Weather clear and dry. Wind S. W. and light. Th: 35°. Stomach empty, clean and healthy. Temperature 100° and a fraction—spirit stationary in five minutes. No gastric juice could be procured. Extracted about half a drachm of fluids, principally mucus.

At 8 o'clock, 45 mins.—Temperature of the stomach 100 1-4° when Thermometer was put three or four inches only into the splenic portion; but rose to 101° when the bulb was let down, eight or nine inches, towards the pyloric extremity.

A circumstance occurred here which I had not noticed before. On settling the stem down into the stomach, a strong contraction of the muscular fibres was indicated, when the bulb had descended near to the pyloric end, by a sudden and peculiar movement of the tube, communicated to the thumb and finger that guided it, and also felt by St. Martin himself. The stomach appeared to contract at that point forcibly, and grasp the bulb, giving it a sudden impulse downwards, so much so as to require a quick compression by the thumb and finger to prevent it from slipping suddenly into the pyloric end. This grasping sensation would continue for half a minute or more, and then appear to relax again. This action occurred every time the bulb passed this point, either up or down. When the bulb was below this point, the spirit rose three fourths of a degree; when raised above, it fell the same. Sometimes the suction motion was stronger than at others, and when the stem was released from the fingers, it would be drawn down towards the pyloric end, its whole length, ten or eleven inches, occasioning considerable distress, vertigo, and a sense of sinking at the scrobiculus cordis.

At 9 o'clock, he breakfasted as yesterday, and kept quiet, most of the time in a recumbent position, on a couch.

At 11 o'clock—Aspects of weather same as in the morning. Th: 46°. Contents of stomach about two thirds diminished. Temperature, 100 3-4,° at three or four inches deep, and a fraction less than 101 1-2°, when sunk to the pyloric extremity, varying proportionably to the length of the stem introduced.

At 12 o'clock, 30 mins., M., stomach nearly empty. Temperature, 101°.

At 1 o'clock, 30 mins., stomach empty. Temperature, 100 3-4°, splenic end--101 1-2°, at pyloric end.

Experiment 80.

At 2 o'clock, P. M., same day, he dined the same as yesterday, on *raw oysters* and *bread*. Temperature of stomach, immediately before eating, 101 1-2°, at pyloric extremity—100 3-4° at splenic end.

At 4 o'clock, 30 mins., stomach half empty. Temperature 101 1-2° at pyloric extremity--rose quick. Took out one drachm of the chyme. Digestion nearly complete—a few particles of bread and oyster to be seen.

At 5 o'clock, 30 mins., stomach nearly empty.—Temperature 101 1-2°, pyloric extremity—rose quick.

At 6 o'clock, stomach still contained alimentary fluids—quite acrid and sharp.

At 6 o'clock, 40 mins., stomach empty.

At 7 o'clock, he supped on *boiled beef and bread*.

Experiment 81.

Jan. 29. At 6 o'clock, A. M.—before rising—Weather clear and dry. Wind N. E. and brisk. Th: 28°. Stomach perfectly healthy, empty and clean. Temperature 100°, at pyloric extremity, and 99° at the other. No gastric secretion. Could not extract ten drops of either gastric juice, mucus or saliva.

At 8 o'clock, 30 mins.—Stomach empty—coats perfectly healthy, and free from any appearance of aphthæ, pustules or red spots, The mucous coat was even and uniform, soft and smooth. Temperature from 100 1-4° to 100 3-4°—rose quick. Extracted three and a half drachms pure gastric juice, containing some flocculi of mucus, but no bile.

At 9 o'clock, he breakfasted on *sausage* and *bread*, and kept exercising--walking smartly for two hours. Returned from walking at 11 o'clock, 30 mins. Stomach two thirds empty. Temperature, 102° and a fraction, at pyloric end, and 101 1-3° at the other—rose quick. 12 o'clock, 30 mins., M., stomach almost empty. Temperature, 101 1-2° and 100 3-4° —rose moderately. 1 o'clock, P. M., stomach empty.

Experiment 82.

At 1 o'clock, 30 mins., P. M., same day, he dined on *stewed oysters* and *bread*, and kept still. 5 o'clock, P. M., stomach empty. Extracted three and a half drachms pure gastric juice. At 6 o'clock, 45 mins., stomach empty. Temperature 101 1-2° and 100 3-4° --rose moderately.

Experiment 83.

Jan. 30. At 6 o'clock, 30 mins., A. M.—Weather cloudy and damp. Wind N. E. and light. Th : 39°. Stomach empty, clean and healthy. Temperature, immediately before rising, 99 1-2° and 98 3-4°—rose slowly.

At 9 o'clock, temperature of stomach, 101 1-2° and 100 3-4°--rose quick. Extracted three drachms gastric juice. It came slowly; the last mixed with yellow bile. He breakfasted on *beef steak*, *bread* and *coffee*.

At 11 o'clock, stomach almost empty. Temperature 101 1-4° and 100 1-2°. When the bulb of the glass tube descended towards the pyloric extremity, the stomach evidently contracted upon it, and drew it forcibly down. If left free to its own motions, the tube would sink to the pylorus, the whole length of the stem, ten or eleven inches, and then rise again of its own accord. When drawn above this point of apparent contraction, into the splenic end of the stomach, towards the perforation, the motion of the bulb was reversed, in a direction towards the fundus of the stomach, not inclining, however to make its exit at the perforation ; but took a sort of irregular motion, revolving the tube from right to left, so as to turn it completely around, in the space of ten or fifteen seconds. This motion was not always present, nor constantly continuous when present; but interrupted, and alternate with the appearance of contraction at the pyloric end ; and distinctly evident only from about one and a half to three hours, or more, after eating, and at the time when the chyme was most rapidly leaving the gastric cavity.

At 12 o'clock, 30 mins., M., he returned from a smart walk—had been all the morning, since breakfast, hard at work, wheeling coal, an unusually severe exercise. Temperature, 102° and 100 3-4°—rose quick. Stomach empty.

Experiment 84.

Jan. 29. At 9 o'clock, 30 mins., A. M.—To two drachms of gastric juice, I put one small, *raw oyster*, weighing one drachm; and to another two drachms of gastric juice, I added one drachm of *stewed oyster*. Set them on the bath, and agitated them frequently.

At 5 o'clock, 30 mins., P. M., the residue of the raw oyster weighed four grains—that of the stewed, weighed eight and three fourths grains. Continued them on bath.

At 9 o'clock, A. M., of the 30th, the raw oyster was completely dissolved; not a particle left, except a trace of dirty brown sediment, the excrementitious part. A grain only of the heart of the stewed oyster, was left, with a trace of the same kind of sediment, as in the raw one. The flavour of the oysters was retained to the last, and even the chymous mass partook of it.

In the article here submitted to the action of the gastric juice, cooking hardens the fibre, and renders it less susceptible of digestion than the raw. This is what we should, *a priori*, judge, from the known properties of this solvent.

Experiment 85.

Jan. 30. At 10 o'clock, A. M., I put 10 grains of *boiled, lean beef*, ten grains of *raw lean beef*, each piece whole and undivided, and ten grains *boiled, lean beef, chopped fine*, into three drachms fresh gastric juice, and placed them on the bath, frequently agitating, as usual.

At 12 o'clock, M., of the 31st, examined and weighed them. The raw piece weighed the same as when first put in—the lean boiled beef weighed eight grains—the chopped, three grains. Added two drachms gastric juice.

At 10 o'clock, A. M., *Feb.* 1st, balance of chopped meat weighed one grain; boiled piece, five grains; raw, ten grains.

Experiment 86.

Jan. 29. At 10 o'clock, A. M., I put three equal parts of *cabbage*, one part *raw*, another *boiled*, and the third, *shaved* fine, (raw) and *macerated in vinegar*, ten grains each, into three drachms of gastric juice, and placed them on the bath.

At 5 o'clock, 30 mins., P. M., I took out and pressed dry the respective parcels. They weighed as follows: the shaved, three and three fourths grains; the raw, five and a half grains; the boiled, six and a quarter grains.

At 10 o'clock, A. M., of the 30th, took out and examined—the raw weighed five and a quarter grains; the shaved, three and a half grains; and the boiled, the same as yesterday evening, six and a quarter grains. I added two drachms gastric juice and continued them on bath.

At 12 o'clock, M., of the 31st, the raw weighed two grains ; the shaved, one and a half grains ; boiled, five grains. Added one drachm of gastric juice.

Feb. 1. Raw, weighed one grain; shaved, one grain ; boiled, two and a half grains.

Experiment 87.

Jan. 30. At 2 o'clock, P. M., he dined on *raw oysters* and *bread*, and kept still. 5 o'clock, 30 mins., stomach empty. Temperature 101 1-4,° and 100 1-2°, spirit rose moderately. 6 o'clock, 45 mins., he supped on *raw oysters* and *bread* ; 10 o'clock, stomach empty.

Experiment 88.

Jan. 31. At 6 o'clock, A. M.—before rising.—Weather rainy. Wind N. E. and light. Thermometer 45°. Stomach empty, clean and healthy. Temperature 100° and 98 1-2°—rose moderately. No fluids in the gastric cavity. Could obtain but half a drachm. The peculiar contraction and relaxation or suction and pulsion motion, were evidently excited this morning by the introduction of the thermometer, but not near so strong as during chymification. When the bulb is sunk down low into the stomach, and suffered to remain there a minute or two, it gives severe pain and distress at the pyloric extremity, like the cramp, or, the sensation frequently described by persons suffering from undigested food in the stomach, and leaves a sense of soreness, if repeated a few times, as was very evident this morning.

At 9 o'clock—temperature of stomach 101 1-2°—he breakfasted on two and a half ounces of *beef steak*, four and a half ounces *soft toast*, and *coffee*.

At 9 o'clock, 30 mins., he laid himself down on his pallet, and I set the Thermometer into his stomach, and continued faithfully and constantly to observe its motions and variations, one hour and forty minutes, until ten minutes past eleven, (two hours and ten minutes after eating.) At first the stomach was full to overflowing, of heterogeneous fluids, in much commotion, as indicated by the movement of the aliment, and of that part of the stem left out of the aperture, nearly four inches.--This commotion continued about half an hour, to ten o'clock. It then seemed to subside ; the general muscular action became less, as indicated by the stem of the thermometer, and motion of the fluids, until half after ten o'clock, when a different motion appeared to commence, indicating considerable forcible contraction upon the bulb of the tube, now about six inches from the aperture towards the pylorus. An irregular turning and twisting of the stem, and a simultaneous downward movement, was succeeded by an apparent relaxation and expulsion motion. These alternate motions and appearances continued to recur every two or three minutes--not uniformly, but at irregular intervals. A sense of distress and uneasiness was felt at the point where the bulb lay, every time these contractions recurred, so as to occasion involuntary manifestations of pain, expressed in the muscular motions of his face. The Thermometer did not perceptibly vary, during all this time, from the usual standard temperature of the interior of the stomach. It was 101 1-4°, at the pyloric extremity, and 100 1-2° in the splenic end, and continued so, during the whole time, ranging between these two points, according as it was moved higher or lower in the gastric cavity. At this time, 11 o'clock, 10 mins., the stomach was about half empty—and chymification rapidly advancing.

At 12 o'clock, 30 mins., M., the stomach was empty and clean. Temperature 101 1-4° and 100 1-2°. Extracted two and a half drachms gastric juice.

Experiment 89.

Feb. 1. At 6 o'clock, A. M.—before rising—Weather clear. Wind N. W. Th: 28°. Stomach empty, clean and healthy. Temperature 100° and 99 1-4°—rose moderately. No gastric juice secreted.

At 8 o'clock—Weather clear, and growing cold. Th: 26°. Temperature of stomach, immediately before going out, 101° and 100°. Returned in 30 minutes. Temperature of stomach the same. Extracted four drachms gastric juice.

At 9 o'clock, he breakfasted on *bread*, *sausage* and *coffee*, and kept exercising. 11 o'clock, 30 mins., stomach two thirds empty. Aspects of weather similar. Th: 29°. Temperature of stomach, 101 1-2° and 100 3-4°. The same appearance of contraction and dilitation, and alternate piston motions were distinctly observed at this examination. 12 o'clock, 30 mins., stomach empty.

Experiment 90.

At 2 o'clock, P. M., same day, he dined on *potatoes* and *meat*. 5 o'clock, weather clear and pleasant. Wind N. W. and light. Th: 32°. Stomach nearly empty. Temperature 102° and 101 1-2°, after walking. 5 o'clock, 30 mins., stomach empty.

Experiment 91.

Feb. 2. At 8 o'clock, 30 mins., St. Martin finished breakfasting on full meal of two and a half ounces *fried sausage*, seven and a half ounces warm *corn bread*, and a pint of *coffee*. Kept gently exercising for one hour, and then increased his exercise to severe walking, two or three miles, for two hours. Stomach full when he started, at 9 o'clock, 45 minutes.

At 12 o'clock, M., returned from walking. Stomach not entirely empty. Oil and bread perceptible. 12 o'clock 30 mins., considerable fluid in the stomach, tinged with yellow bile. No distinct particles of food to be distinguished. 1 o'clock, P. M., stomach empty and clean. Extracted two drachms pure gastric juice.

Severe exercise, in this instance, is supposed to

have retarded digestion, as well as the peculiar kind of food eaten.

Experiment 92.

At 1 o'clock, 30 mins., P. M., same day, he dined on four ounces of *fresh*, *boiled beef*, (cold) and five ounces *bread*, and continued walking smartly, for three and a half hours, till 4 o'clock, 45 mins., P. M. Stomach nearly empty. Dinner almost completely chymified. 5 o'clock, stomach empty.

Experiment 93.

Feb. 3. At 8 o'clock 45 mins., extracted four drachms gastric juice. He breakfasted on full meal, two and a half ounces *boiled beef*, seven and a half ounces *bread*, and one pint of *coffee*; and kept perfectly still. 12 o'clock, 30 mins., M., stomach not empty. 1 o'clock, P. M., stomach empty and clean. Extracted one and a half drachms gastric juice.

This indicates that a complete state of repose is unfavourable to speedy digestion.

Experiment 94.

At 1 o'clock, 30 mins., P. M., same day, he dined on four ounces *fresh*, *boild beef*, five ounces of *bread*, and *patatoes*. 6 o'clock, stomach not entirely empty; but none of the meat remaining. 6 o'clock 15 mins., very little of the bread and potatoes to be seen. 6 o'clock, 30 mins., stomach empty.

Experiment 95.

Feb. 3. At 12 o'clock, M., I put two equal and entire pieces of *parsnip*, ten grains each, one boiled, and the other raw; the same kinds and quantities of *carrot*; and the same of *potato*—into four drachms of gastric juice, and placed them on the bath.

At 12 o'clock, M., on the 4th, the vegetables taken out and wiped or filtered as dry as when put in, shewed the following result:

The piece of raw parsnip, weighed three grains; the boiled, one grain. Raw carrot, three and one fourth grains; boiled, half a grain. Raw potato, eight and a half grains; boiled, no entire particle could be distinguished—a fibrous and farinaceous residuum of six grains remained on the filter.

At 12 o'clock, M., on the 5th, the pieces of parsnip and carrot were almost entirely dissolved, a grain or two of the raw carrot, and fibrous centre of the parsnip, only remaining. About a grain of the roughish white farina of the boiled potato, remained. The raw potato was a little softened and wasted on the surface, but weighed the same as at last examination, eight and a half grains.

This is an illustration of the necessity of tenderness and susceptibility of division of the articles of diet, for speedy solution by the gastric juice. The raw potato retained nearly its weight, after the other articles were dissolved.

Experiment 96.

At 3 o'clock, P. M., same day, I took two equal quantities, two drachms each, of saliva, acidulated to about the flavour of gastric juice—one with acetic, and the other with muriatic acid—and put into each, two pieces of parsnip, and two of carrot, one of each boiled, and the other raw; each weighing ten grains; and placed them on the bath.

At 3 o'clock, P. M., on the 4th, the carrot in the saliva and muriatic acid, had lost nothing—the parsnip, only two grains. In the acetous menstruum, both kinds remained the same as when put in. The fluids of both were unaltered in their sensible qualities and appearances.

After continuing them on the bath, with frequent agitation, for twenty-four hours longer, the parsnip, in the muriatic menstruum, had lost four grains—the carrot nothing. The parsnip in the acetic mixture, had lost six grains, and the carrot four grains, but appeared to have been rather macerated and diffused, than dissolved or digested.

I now mixed them all together, and continued them on the bath, for twenty-four hours longer; at the end of which time, the whole remaining mass of vegetable matter, weighed twelve grains. The fluid appeared now a little more chymous, and was rather turbid.

This is an example of a species of solution, performed by chemical agents, having some resemblance to digestion. It is not at all probable, however, that this mixture was in a state of preparation for the action of the pancreatic and hepatic fluids; but if placed in the stomach, would require the same action of the gastric juice, as other diet would.

Experiment 97.

Feb. 4. At 9 o'clock, A. M., he breakfasted on two and a half ounces of *boiled beef*, six ounces of *bread*, and one pint of *coffee*. Exercised smartly for three hours. At 12 o'clock, 30 mins., M., chymification complete. Stomach empty.

Experiment 98.

Feb. 5. At 9 o'clock, A. M., he breakfasted same as yesterday, and kept still. 11 o'clock, stomach nearly full. 12 o'clock, M., considerable yet in the stomach; oil and bread plainly to be seen. 12 o'clock, 30 mins., contents of stomach not yet gone. 1 o'clock, P. M., stomach almost empty. 1 o'clock, 15 mins., stomach empty.

Experiment 99.

Feb. 7. At 8 o'clock, 30 mins., A. M., I put twenty grains *boiled codfish* into three drachms gastric juice, and placed them on the bath.

At 1 o'clock, 30 mins., P. M., fish in the gastric juice, on the bath, was almost dissolved, four grains only remaining—fluid opaque, white, nearly the colour of milk. 2 o'clock, the fish in the vial, all completely dissolved.

Experiment 100.

Feb. 7. At 9 o'clock, A. M., breakfasted on *boiled codfish* and *bread.* Digested in four hours and a half.

Experiment 101.

Feb. 8. At 10 o'clock, 30 mins., A. M., I put two parcels, ten grains each, of *strong cheese*, one masticated, and the other an entire piece, into three drachms gastric juice. At 6 o'clock, P. M., the masticated portion was all completely digested, scarcely a trace left on the filter. The entire piece had lost four and three fourths grains—five and one fourth grains remaining undissolved, and of the same shape as when put in, having lost its superfices only. This piece continued gradually to diminish, for twenty four hours, when it was completely dissolved.

Experiment 102.

Feb. 12. At 1 o'clock, 30 mins., P. M., he dined on *mutton*, and *barley soup* and *bread.* Digested in three and a quarter hours.

Experiment 103.

Feb. 13. At 2 o'clock, 15 minutes P. M., he dined on *mutton* and *barley soup* and *bread.* Digested in three and a quarter hours.

Experiment 104.

Feb. 14. At 9 o'clock, A. M., I took forty grains *masticated*, *broiled beef steak*, divided into two equal parts—put one into four drachms gastric juice, and the other, into four drachms of a mixture of dilute *muriatic* and *acetic* acids, reduced with water to the flavor of the gastric fluid, as nearly as practicable—

three parts of the muriatic to one part of the acetic, Placed them together on the bath. At 6 o'clock, P, M., the meat in the gastric juice was all dissolved; that in the dilute acids, when filtered, left a residuum of nine grains, of a jelly-like consistence. The fluids, also, differed in appearance. That from the gastric juice was opaque, and of a lightish grey colour, depositing a brown sediment on standing. The other was also opaque, and of a reddish brown colour, but deposited no sediment.

This was an attempt to imitate the gastric juice. It was not satisfactory. Probably the gastric juice contains some principles inappreciable to the senses, or to chemical tests, besides the alkaline substances already discovered in it.

Experiment 105.

At the same time of the above experiment, (104) I put the same quantities of pure dry *gelatine*, (ichthyocolla) into exactly similar quantities and kinds of fluids, and placed them all together on the bath.

At 6 o'clock, P. M., the gelatine in the gastric juice was all completely dissolved—that in the dilute acids, after being placed on the filter, left a residuum of three grains of a jelly-like substance. These two fluids differed in appearance. That from the gastric liquor was of an opaque, whitish colour, with little fine, brown sediment—that from the acid menstruum was also opaque, but of a reddish brown colour, and of a thin, mucilaginous consistence, with no sediment.

One drachm of infusion of nutgalls, added to the gastric solution, immediately afforded a rich, creamlike fluid, and slowly precipitated a fine, compact sediment. The same quantity of infusion of galls, added to the other, immediately formed the whole mass into a coarse, brown coagulum. After standing a while, it afforded a large, loose, brownish sediment, and a light coloured fluid, which, on standing, became white as milk; and the sediment became compact and remained so.

The precipitates, after the addition of the *tan* taken out and filtered, weighed as follows—that in the gastric solution, eighteen grains; the other, forty grains—the difference of weight being about equal to the quantity of gelatine put in.

Experiment 106.

Feb. 15. At 9 o'clock, 45 mins., A. M., repeated the last (105th) experiment, with *gelatine*, and the gastric juice, and dilute acids, in the same proportions.

At 3 o'clock, 15 mins., P. M., the gelatine in gastric juice, all dissolved, to a mere mite—that in acid mixture, nearly so, six grains only, remaining on the filter, of a jelly-like consistence. The fluid of the gastric portion had a bluish white colour, and the other, yellowish, or about the colour of dry gelatine.

At 6 o'clock, the gelatine in the acid menstruum all dissolved. Fluids of both, nearly similar.

One drachm infusion of nutgalls, added to each, instantly formed loose, lightish coloured coagulæ in both; threw down a compact sediment in the gastric solution, and left an opaque, milky fluid. The coarse coagulæ in the acid menstruum, continued suspended throughout the mass of fluids, for a long time, gradually subsiding. At the end of forty eight hours, it had become precipitated to the bottom into a compact mass, and exhibited distinct particles of the entire, undissolved gelatine, mixed with a dirty white coloured, curd-like substance.

Experiment 107.

Feb. 15. At 1 o'clock 30 mins., P. M., he dined on *boiled codfish* and *bread.* Digested in four hours and a half.

Experiment 108.

Feb. 16. At 1 o'clock, 45 mins., P. M., he dined on *mutton soup* and *bread*—6 o'clock stomach empty. Digested in four hours and a quarter.

Experiment 109.

Feb. 19. At 9 o'clock, A. M., I put twenty grains of *boiled fat pork*, cut fine, into three drachms of *clear* gastric juice, and the same kind and quantity into three drachms of gastric juice, strongly tinged with yellow *bile*, with a view to ascertain whether there be any difference in their solvent effects upon fat meats. Placed both on axilla. At 1 o'clock, P. M., the pork in the gastric juice, tinged with *bile*, dissolved to less than one grain—that remaining undissolved, in the *clear* juice, weighed two grains and a half.

Experiment 110.

Feb. 20. At 1 o'clock, 30 mins., P. M., I put three parcels, ten grains each, of *boiled codfish*, into three separate portions of gastric juice, one *pure*, another containing *bile*, and the third, a *clear*, *limpid*, *slightly acid fluid*, taken from the stomach after active exercise and profuse perspiration, in more abundant quantity than usual. Placed them all on the bath.

At 1 o'clock, 30 mins., P. M., of the 21st, I took out and weighed the three parcels of fish. The result was as follows: that in the pure gastric juice weighed two and a half grains; that in the yellow, three grains; and the other, six grains.

This shows that other than oily food is retarded by the admixture of bile in the gastric juice.

Experiment 111.

Feb. 23. At 9 o'clock, 45 mins., P. M., I took out two parcels, one drachm each, of gastric juice, one pure, and the other of the clear, limpid fluid, extracted under the circumstances mentioned in the last experiment, and put eight grains of *lean beef*, finely cut, into each; and placed them on the bath together. After being treated alike on the bath for six or eight hours, the residuum in the pure gastric juice, weighed three grains; and that in the limpid fluid, six grains and a half.

Experiment 112.

Feb. 24. At 9 o'clock, 30 mins., A M., having extracted gastric juice, containing a large proportion of yellow bile, I put twenty grains of *strong cheese*, cut small, into two drachms of it; and the same quantity and kind of cheese, into two drachms of pure gastric juice: placed them together on the bath.

At 9 o'clock, P. M., residuum in the yellow juice, weighed five grains; that in the clear juice six grains. Returned them to the bath.

At 9 o'clock, A. M., of the 25th, the cheese in the yellow juice, all dissolved; of that in the clear juice, two grains remained.

Experiment 113.

March 26. At 8 o'clock, 15 mins.--Weather clear. Stomach empty and healthy. Introduced Thermometer (Pool's glass) three fourths the length of its tube, eight or ten inches, and continued it five minutes. Spirit stationary at 100 1-2°.

At 9 o'clock, suspended, through the aperture, into the stomach, enclosed in a muslin bag, forty grains of *broiled, fresh codfish*, previously masticated, and imbued with saliva; and he immediately afterwards breakfasted on the same kind of fish, a small quantity of *bread*, and *coffee*, and kept exercising moderately.

At 11 o'clock, stomach full of fluids. 2 o'clock, P. M., chymification complete. Bag empty.

Experiment 114.

March 27. At 9 o'clock, 15 mins., A.M., he breakfasted on *fresh, broiled fish* (*Flounder*) *bread* and *coffee*, and kept exercising moderately. 11 o'clock, stomach half empty—pulp of bread only appeared. 11 o'clock, 30 mins., particles of fish and bread still to be seen in the stomach. 1 o'clock, P. M,. stomach entirely clear of food. Temperature 101°.

Experiment 115.

I took *dilute muriatic* acid, reduced it to the strength and taste of the gastric juice, as nearly as practicable, three drachms ; *dilute acetous acid*, to about the same flavour, one drachm—mixed them together, and put into this mixture, one scruple of *broiled steak*, cut fine ; and the same quantity and kind of meat into four drachms of gastric juice. Placed them both on the bath. In six hours and three quarters, the meat in the gastric juice, taken out and filtered, weighed two grains only—that in the acid mixture, treated in the same way, was not dissolved ; but had lost its fibrous form ; and was converted into a tremulous, jelly-like mass, too tenacious to pass through the filter, and weighed more than when first put in. It did not appear like chyme, nor resemble that in the gastric juice.

After digesting eight hours longer, on the bath, the contents of the acid mixture had become nearly dissolved or diffused, and when run through the filter, left only a very little of the jelly-like mass, so abundant in the first examination. The liquid was now more like, though not exactly similar, to that of the gastric portion ; this being opaque, and of a lightish grey colour, affording a dark brown sediment on standing ; that from the acid menstruum, was also opaque, of reddish brown colour, but deposited no sediment.

Two drachms of the infusion of nutgalls, added to the gastric portion, threw down a fine, reddish brown precipitate, and afforded an opaque fluid, of similar colour. Two drachms of the infusion, added to the acid mixture, threw down a more copious precipitate, and left a clearer and thinner fluid, of a yellowish colour, and nearly transparent.

Experiment 116.

A drachm of the concentrated, disinfecting solution of chloride of soda, prepared according to the formula of Labarraque, was added to a drachm of an extremely putrid mixture of beef, macerated in water—the putridity speedily disappeared; but not more so than when a drachm of *pure gastric juice* was added to a similar quantity of the same putrid mixture.

MICROSCOPIC EXAMINATIONS.

The following Microscopic examinations, were made with Jones' compound Microscope, in presence of Professor Dunglison and of Capt. H. Smith, of the Army. They afford, however, very little information on the subject of digestion, and show that no very satisfactory results are attainable from Microscopic examinations of Chyme.

I. *Pure gastric juice*, exhibited the appearance of water, except that there were perceptible, a very few minute globules.

II. The chymous product of the *gastric juice and unmasticated, lean beef*, exhibited globules of various sizes, resembling those of the blood, having a transparent centre, and opaque margin, with various very fine filaments, of apparently undigested fibrine.

III. Product of *gastric juice and albumen*—exhibited appearances resembling, considerably, those presented by the gastric juice alone—no distinct globular arrangement.

IV. Chyme from *gastric juice and tendon of veal*—exhibited numerous minute, apparently fleshy, particles—no globular appearance.

V. Chyme from *gastric juice* and *fowl* and *bread* --In the comparatively clear portion, (taken without shaking the vial) exhibited a few undissolved particles, and very few globules. A portion taken after shaking the vial, exhibited considerably more particles, and a greater number of globules.

VI. The product of the same kind of aliment, (*fowl* and *bread*) macerated in water, exhibited numerous undissolved particles, with few globules;—the globules not so regularly formed, as in the foregoing experiment.

VII. Product of *gastric juice*, and *soup, made from fresh beef*, exhibited globules extremely numerous, and distinctly formed, far more so, than in any of the preceding experiments—and a few particles of meat.

VIII. *Impure gastric juice*, or that with an *admixture of green bile*, when taken from the stomach, exhibited numerous amorphous particles, with few globules.

IX. Chyme, *artificially formed*, from *pork* and *bread*, exhibited numerous globules of different sizes, apparently oily.

X. *Chymous product* of *gastric juice* and *fat pork, formed in the stomach*, exhibited a beautiful appearance of large transparent globules, of different sizes, evidently oily.

XI. *Fat pork, macerated in pure water*, presented appearances of globules precisely similar to those in the products of digestion.

EXPERIMENTS, &c.

FOURTH SERIES.

PLATTSBURGH, N. Y. 1833.

The following Gastric Experiments, and Examinations of the stomach, have been made since the manuscript of the previous part of this work was prepared for the press.

EXAMINATIONS OF THE TEMPERATURE AND APPEARANCE OF THE INTERIOR OF THE STOMACH.

I. *July* 9. 6 o'clock, A. M. Weather cloudy and damp. Wind W., light. Stomach empty and clean. Introduced glass thermometer, at the aperture, bulb nine inches down towards the pylorus—temperature 100°, Fahrenheit, before rising from his bed.

II. *July* 10. 6 o'clock, A. M. Weather clear. Wind W., brisk. Th: 63°. Stomach empty and clean. Temperature 100°, before rising. 9 o'clock, P. M. Weather clear and calm. Th: 75°. Stomach empty. Temperature 101°, after moderate exercise in open air.

III. *July* 11. 6 o'clock, A. M. Weather cloudy. Wind N. E., brisk. Th: 65°. Stomach empty and clean. Temperature 100° before rising 8 o'clock,

30 mins. Weather clear and dry. Wind S., brisk. Temperature of stomach 101°, after exercise. 9 o'clock 30 mins., P. M. Weather hazy. Wind S. W. light. Th: 75°. Temperature 101°.

IV. *July* 12. 6 o'clock, A. M. Weather clear. Wind W., brisk. Th: 70°. Stomach empty. Temperature 100 1-2°, after going out into the open air. 9 o'clock, P. M. Weather clear. Wind W., light, Th: 76°. Temperature 101 3-4°. Stomach empty.

V. *July* 13. 5 o'clock, 30 mins., A. M. Weather clear, serene and calm. Thermometer 69°. Stomach empty, healthy and clean. Temperature 99 1-2°, before rising from his bed. 6 o'clock, 30 mins. Weather same as at last examination. Stomach empty. Temperature 100 3-4°, after rising and walking out in open air, twenty or thirty minutes. 6 o'clock, 45 mins. Returned from a smart walk, exercising so as to produce gentle perspiration. Temperature 101 3-4°.

VI. *July* 14. 5 o'clock, 30 mins., A. M. Weather variable—heavy thunder shower, during the night. Wind S., moderate. Th: 75°. Stomach empty.—Temperature 100° on rising from bed—100 3-4° after walking out into the open air, and immediately back. 9 o'clock, P. M. Weather rainy—atmosphere oppressive. Th: 79°. Wind S., light. Temperature of stomach 102°. St. Martin has been in the woods all day, picking whortleberries, and has eaten no other food since 7 o'clock in the morning, till 8 at evening. Stomach full of berries and chymifying aliment, frothing and foaming like fermenting beer or cider—appears to have been drinking liquor too freely.

VII. *July* 15. 5 o'clock, 30 mins., A. M. Weather clear. Wind W., light—air damp—ground wet. Th: 74°. Stomach empty. Temperature 100°, before rising. 7 o'clock, 30 mins. Weather, wind, &c. same as at last examination. Th: 74°. Stomach empty. Temperature 102°, immediately after smart exercise. 1 o'clock, 30 mins., P. M. Weather clear and pleasant, since 8 o'clock, (till within fifteen minutes, in which interim, has fallen a light shower of rain.) Wind W., light. Th: 74°. Stomach empty. Temperature 100 1-2°—has been at manual exercise for four hours. 9 o'clock, P. M.—Weather and wind, same. Th: 72°. Temperature 101 3-4°. Stomach full of chymous fluid, oil, and pulp of bread and cakes, eaten for supper, two hours previous to examination.

VIII. *July* 16. 7 o'clock, 30 mins., A. M. Weather cloudy. Wind W., light. Th: 73°. Stomach empty. Temperature 101°, after rising and before exercising. 9 o'clock, P. M. Weather cloudy, damp and chilly. Th: 70°. Temperature 101 1-2°.

IX. *July* 28. 9 o'clock, A. M. Weather clear. Wind N. W., brisk. Th: 66°. Stomach empty—not healthy—some erythema and aphthous patches on the mucous surface. St. Martin has been drinking ardent spirits, pretty freely, for eight or ten days past—complains of no pain, nor shows symptoms of any general indisposition—says he feels well and has a good appetite.

X. *August* 1. 8 o'clock, A. M. Examined stomach before eating any thing—inner membrane morbid—considerable erythema and some aphthous patches on the exposed surface—secretions vitiated—extracted about half an ounce of gastric juice—not clear and pure as in health—quite viscid.

XI. *August* 2. 8 o'clock, A. M. Circumstances and appearances very similar to those of yesterday morning. Extracted one ounce of gastric fluids—consisting of unusual proportions of vitiated mucus, saliva, and some bile, tinged slightly with blood, appearing to exude from the surface of the erythema, and aphthous patches, which were tenderer and more irritable than usual. St. Martin complains of no sense of pain, symptoms of indisposition, or even of impaired appetite. Temperature of stomach 101°.

XII. *August* 3. 7 o'clock, A. M. Inner membrane of stomach unusually morbid—the erythematous appearance more extensive, and spots more livid than usual; from the surface of some of which, exuded small drops of grumous blood—the aphthous patches larger and more numerous—the mucous covering, thicker than common, and the gastric secretions much more vitiated. The gastric fluids extracted this morning were mixed with a large proportion of thick ropy mucus, and considerable muco-purulent matter, slightly tinged with blood, resembling the discharge from the bowels in some cases of chronic dysentery. Notwithstanding this diseased appearance of the stomach, no very essential aberration of its functions was manifested. St. Martin complains of no symptoms indicating any general derangement of the system, except an uneasy sensation, and a tenderness at the pit of the stomach, and some vertigo, with dimness and yellowness of vision, on stooping down and rising again—has a thin, yellowish brown coat on his tongue, and his countenance is rather sallow—pulse, uniform and regular; appetite good; rests quietly, and sleeps as well as usual.

XIII. *August* 4. 8 o'clock, A. M., stomach empty; less of those aphthous patches than yesterday; erythematous appearance more extensively diffused over the inner coats, and the surface inclined to bleed; secretions vitiated. Extracted about an ounce of gastric fluids, consisting of ropy mucus, some bile, and less of the muco-purulent matter, than yesterday; flavour peculiarly fœtid and disagreeable; alkalescent and insipid; no perceptible acid; appetite good; rests well, and no indications of general disease or indisposition.

XIV. *August* 5. 8 o'clock, A. M., stomach empty; coats less morbid than yesterday; aphthous patches mostly disappeared; mucous surface more uniform, soft, and nearly of the natural, healthy colour; secretions less vitiated. Extracted two ounces gastric juice, more clear and pure, than that taken for four or five days last past, and slightly acid: but containing a larger proportion of mucus, and more opaque than usual, in a healthy condition.

XV. *August* 6. 8 o'clock, A. M., stomach empty; coats clean and healthy as usual; secretions less vitiated. Extracted two ounces gastric juice, of more natural and healthy appearance, with the usual gastric acid flavour; complains of no uneasy sensations, or the slightest symptom of indisposition; says he feels perfectly well, and has a voracious appetite; but not permitted to indulge it to satiety.—He has been restricted from full, and confined to low diet, and simple, diluent drinks, for the last few days, and has not been allowed to taste of any stimulating liquors, or to indulge in excesses of any kind.

Diseased appearances, similar to those mentioned above, have frequently presented themselves, in the course of my experiments and examinations, as the reader will have perceived. They have generally, but not always, succeeded to some appreciable cause. Improper indulgence in eating and drinking, has been the most common precursor of these diseased conditions of the coats of the stomach. The free use of ardent spirits, wine, beer, or any intoxicating liquor, when continued for some days, has invariably produced these morbid changes. Eating voraciously, or to excess; swallowing food coarsely masticated, or too fast; the introduction of solid pieces of meat, suspended by cords, into the stomach; or of muslin bags of aliment, secured in the same way; almost invariably produce similar effects, if repeated a number of times in close succession.

These morbid changes and conditions are, however, seldom indicated by any ordinary symptoms, or particular sensations described or complained of, unless when in considerable excess, or when there have been corresponding symptoms of a general affection of the system. They could not, in fact, in most cases, have been anticipated from any external symptoms; and their existence was only ascertained by actual, ocular demonstration.

It is interesting to observe to what extent the stomach, perhaps the most important organ of the *animal* system, may become diseased, without manifesting any external symptoms of such disease, or any evident signs of functional aberration. Vitiated secretions may also take place, and continue for some time, without affecting the health, in any *sensible* degree.

Extensive active or chronic disease may exist in the membranous tissues of the stomach and bowels, more frequently than has been generally believed; and it is possible that there are good grounds for the opinion advanced by a celebrated teacher of medicine, that most febrile complaints are the effects of gastric and enteric inflammations. In the case of the subject of these experiments, inflammation certainly does exist, to a considerable extent, even in an *apparent* state of health—greater than could have been believed to comport with the due operations of the gastric functions.

Experiments, &c.

Experiment 1.

Sept. 18. At 8 o'clock, 45 mins., A. M., St. Martin breakfasted on four ounces of *fresh salmon trout, fried,* three ounces of *bread,* and drank half a pint of *water.* The coats of the stomach were not perfectly healthy; some aphthous patches and dark red spots to be seen on the mucous surface; gastric juice slightly viscid; acid taste distinctly perceptible. At 10 o'clock, 15 mins., stomach entirely empty. Breakfast completely chymified and gone; nothing but a little gastric juice and flocculi of mucus, remaining in the stomach.

Experiment 2.

Sept. 18. At 2 o'clock, P. M., he dined on six ounces of *boiled, fresh, salmon trout*, three ounces of *bread*, and a *potato*, and drank half a pint of *water*. Continued at work, sawing and splitting wood. He had eaten nothing from the time he took his breakfast; had been hard at work all the time; looked, and said he felt quite fatigued.

At 3 o'clock, 40 mins., stomach about half full of a nearly homogeneous semi-fluid, of a rich milk or cream colour, and about the consistence of fine corn meal gruel—a few small particles of the fish, and some of the potato, could be distinguished. 4 o'clock, 15 mins., stomach empty and clean.

Experiment 3.

Sept. 20. At 1 o'clock, 15 mins., P. M., he dined on three ounces *fat pork*, and one pint of *corn* and *beans*, (green,) two ounces of *bread*, and half a pint of *water*; and kept exercising. Digested in three hours and three quarters.

Experiment 4.

Sept. 21. At 8 o'clock, A. M., he breakfasted on eight ounces of *beef's liver, broiled*, two ounces of *bread*, and drank half a pint of *water*. Continued usual exercise. 9 o'clock 30 mins., stomach full of partially chymified food, considerable oil, (melted butter,) floating on the surface; black pepper mingled with it, and emitting a strong aromatic odour of the spice. 10 o'clock, 30 mins., stomach empty and clean. Extracted two drachms of gastric juice.

Experiment 5.

At 1 o'clock, 30 mins., P. M., same day, St. Martin dined on one pint of rich *beef* and *vegetable soup*, made of the joint, marrow bone and muscle of the leg of an ox, three ounces of *bread*, and continued moderate exercise. 3 o'clock, 15 mins., stomach nearly full of thick, greyish white, porridge-like semi-fluid.

with a thick pellicle of oil floating on the surface 4 o'clock, P. M., stomach empty.

Experiment 6.

Sept. 30. At 7 o'clock, 30 mins., A. M., he breakfasted on *bread* and *milk*, and continued his usual exercise. 8 o'clock, 30 mins., stomach nearly full of milky fluid, pulp of bread and coagulæ. 9 o'clock, contents of stomach considerably diminished since last examination—took out a portion, nearly chymified; very little fine coagulæ perceptible; bread in small particles, reduced to a greyish, soft pulp; the menstruum of a whitish, whey-colour and consistence. 9 o'clock, 30 mins., chymification complete. Stomach empty and clean.

The portion taken out of the stomach, at 9 o'clock, into a vial and continued in the axilla, till 12 o'clock, M., was almost completely chymified; small pulpous particles of bread only, discernible; the fluid of a rich whey, or gruel colour and consistence; a little loose, light coloured sediment fell to the bottom, on standing.

Experiment 7.

Oct. 1. At 1 o'clock, 30 mins., P. M., St. Martin dined on *boiled*, *fresh*, *lean beef*, *potatoes* and *bread*, and continued his usual exercise. 4 o'clock, 15 mins., stomach empty.

Experiment 8.

Oct. 2. At 1 o'clock, 30 mins., P. M., he dined on same kind of food as yesterday, *lean*, *boiled beef*, *potatoes* and *bread*, dressed with a liberal quantity of strong mustard and vinegar, and continued the same exercise. 3 o'clock, 30 mins., stomach nearly full of heterogeneous mixture. 4 o'clock, 30 mins., stomach still contains chyme and some undissolved food; fluids taste and smell quite strongly of the mustard; complains of more smarting at the edges of the aperture, than usual; some slight morbid appearance on the mucous surface. 5 o'clock, stomach empty.

These two last experiments were made under almost exactly similar conditions of the stomach, with a view to notice the effects of this kind of stimulating condiment. The result was, that it apparently retarded the process of digestion; no other appreciable cause existed, for this difference of result. The stomach presented the usual healthy appearance immediately previous to the ingestion of the meal. Nothing occurred to interfere with, or interrupt the digestive functions. The slight morbid appearance on the mucous surface, towards the close of chymification, I conceive to have been more the effect of the over excitement of the mustard than any other cause.

It would seem then, that stimulating condiments, instead of being used with impunity, are actually prejudicial to the healthy stomach. They can only be required, and taken with benefit, when the gastric apparatus is languid and relaxed, and requires stimulants to excite the tone and action of its vascular tissues.

Experiment 9.

Oct. 3. At 2 o'clock, 35 mins., P. M., St. Martin ate nine ounces of *raw, ripe, sour apples.* 3 o'clock, 30 mins., stomach full of fluid and pulp of apples; quite acrid, and irritating the edges of the aperture, as is always the case when he eats acescent fruits or vegetables. 4 o'clock, stomach not empty; contents more sharp and acrid; pulp of apple still to be seen. 4 o'clock, 40 mins., stomach empty; morbid appearance of the gastric surface considerably increased.

Experiment 10.

Oct. 7. At 8 o'clock, A. M., he breakfasted on *bean soup,* made with *fresh beef* and *bread.* Digested in three hours. And at 2 o'clock, P. M., he dined on the same, which was digested in three and a quarter hours.

Experiment 11.

Oct. 10. At 8 o'clock, A. M. Weather fair.—Wind W., light. Th: 61°. Stomach empty and healthy. Temperature 101°, after moderate exercise. Breakfasted on *baked potatoes* and *bread.* 10 o'clock, stomach nearly empty; a little chymous fluid to be seen; quite acrid. Temperature 101½°, after usual exercise. 10 o'clock, 45 mins., stomach empty. Temperature 101½°.

Experiment 12.

At 2 o'clock, P. M., same day. Weather hazy. Wind S., moderate. Th: 61°. Stomach empty and healthy. Temperature 101 3-4°, after exercise. Dined on *roast beef*, *bread*, *potatoes* and *boiled cabbage.* 4 o'clock. Wind S. W., brisk—raining. Th: 61°. Stomach half full of heterogeneous mass of acrid fluid, oil, beef and cabbage. Temperature 103°; had been smartly exercising for two hours. 7 o'clock, 30 mins., wind and weather same as at 4 o'clock.—Th: 63°. Stomach empty. Temperature 102.—Exercise continued moderately till this examination.

In this experiment,. the temperature of the stomach rose to 103°, one degree higher than I have ever before observed it to rise; and chymification was protracted.

Whether these two circumstances were occasioned by unusually increased exercise, and the consequent fatigue of the system, or from the nature of the aliment eaten, and the unusual fulness of the meal, I am not able positively to say; but am inclined to think, from previous observations, that they are attributable to the latter—as the usual morbid appearances, consequent on too full alimentation, followed this meal in the course of twenty four or thirty six hours—as may be seen by the two subsequent experiments.

Experiment 13.

Oct. 11. 7 o'clock, 30 mins., A. M. Weather fair.

Wind N. W., brisk. Th: 32°. Stomach empty.—Temperature 100½°, after moderate exercise in open air. 8 o'clock, 45 mins., wind and weather, same. Th: 38. Stomach empty. Temperature 102°—had been smartly exercising, shovelling dirt, for an hour or more, and was quite warm. Breakfasted on *stewed veal* and *bread*. 11 o'clock, stomach not empty. Temperature 102°—continues exercise. 12 o'clock, stomach contains a very little chymous fluid, and a trace of the muscular fibres of the veal. 12 o'clock, 30 mins., stomach empty.

Experiment 14.

At 2 o'clock, P. M., same day, he dined on *fried veal* and *bread*, and continued moderate exercise.—6 o'clock, 30 mins., stomach empty. Temperature 101 3-4°. Some morbid appearance on the mucous surface.

At 8 o'clock, 30 mins., weather fair and calm. Th: 36°. Stomach empty; slightly morbid, with few aphthous spots. Temperature 101½°; had been still and quiet for three or four hours.

Experiment 15.

Oct. 12. At 7 o'clock, 30 mins., A. M. Weather hazy. Wind S., light. Th: 36°. Stomach empty—coats not entirely healthy—some erythema and aphthous patches. Temperature 101°, after usual morning exercise. 8 o'clock—circumstances same as at last examination. Temperature 101°. Breakfasted on *fresh beef*, *fried dry*, and *bread*. 10 o'clock, stomach full of fluids; particles of beef, bread and oil, distinctly to be seen. Temperature 101°. 12 o'clock, stomach empty.

Experiment 16.

Oct. 13. At 7 o'clock, A. M. Weather rainy. W., N. E., brisk. Th: 42°. Stomach empty. Temperature 101°, after morning exercise. 9 o'clock, temperature same. Breakfasted on *old*, *salted pork*, *fat* and *lean* together, (*fried*) four ounces of *bread*, and the *yolks* of *six eggs*, fried hard with the pork. 11 o'clock, contents of the stomach heterogeneous; distinct particles of lean pork, egg and oil to be seen; fluid sharp and acrid. Temperature 101°. 12 o'clock, M., oil and egg still to be seen, floating in a milky, chymous fluid; the oil, or lard on the surface, and the egg, in firm coagulæ, diffused through the fluid. Temperature 101°. 1 o'clock, 15 mins., P. M., stomach empty and clean. Temperature 101°—was quiet and inactive during this experiment.

Experiment 17.

At 2 o'clock 20 mins., P. M. same day, St. Martin dined on six ounces of the *spinal marrow* of an ox, steam-cooked, and seasoned with a little *butter*, *vinegar*, *salt* and *pepper*, and three ounces of *bread*. 4 o'clock, P. M., contents of stomach a perfectly milk-white, semi-fluid pulp. Temperature 102°. 5 o'clock, 10 mins., stomach empty and clean.

Experiment 18.

At 6 o'clock P. M., he ate a full meal of *boiled rice*, simply cooked in water, and seasoned with a little salt. 7 o'clock, stomach empty and clean; not a vestige of the rice to be seen.

Experiment 19.

Oct. 14. At 9 o'clock, A. M., he breakfasted on the *albumen* of *six eggs*, *fried hard*, in pork fat. 12 o'clock 15 mins., M., all chymified—stomach empty.

Experiment 20.

At 1 o'clock, P. M., same day, he dined on eight ounces, *boiled beef's brains*, seasoned with *salt*, and a small piece of *bread*. 2 o'clock, stomach full of milk-white, pulpous, or porridge-like semi-fluid; slightly acid taste, and of a bland, insipid flavour. 2 o'clock, 30 mins., stomach almost empty; scarcely any of the white, pulpous mass to be seen. Temperature 102.° 3 o'clock 15 mins.. P. M., stomach empty and clean.

Experiment 21.

At 3 o'clock 30 mins., P. M., same day, St. Martin ate a small head of *raw cabbage*, weighing ten ounces. 5 o'clock, 45 mins., not a particle of the cabbage in the stomach; little albuminous, or greyish, chymous fluid, only remained.

Experiment 22.

At 6 o'clock, 30 mins., P. M., he ate six ounces *boiled leg of fresh mutton*, rare done, dressed with a little *melted butter* and *vinegar*, and two ounces of *bread*. 8 o'clock, stomach empty and clean.

Experiment 23.

Oct. 15. At 8 o'clock, 45 mins., breakfasted on *three fresh eggs*, softly coagulated, by being broken and put raw into boiling water, and three ounces of dry *bread*. 12 o'clock, M. stomach empty.

Experiment 24.

At 1 o'clock, 30 mins., P. M., he dined on *apple dumplings*, made of wheaten dough and sweet apples, boiled, one and a half pounds. 4 o'clock, all chymified, and stomach empty.

Experiment 25.

Oct. 16. At 8 o'clock, 45 mins., A. M., he breakfasted on *broiled, salted pork* and *bread*. 12 o'clock, M., all chymified, and gone from the stomach.

Experiment 26.

At 1 o'clock, P. M., same day. he dined on *raw, salted pork*, cut thin, and eaten with dry *bread*. Digested in three hours.

Experiment 27.

At 4 o'clock, 30 mins., same day, he ate half a pound of *raw cabbage*, cut fine, and macerated in *vinegar*. 5 o'clock 45 mins., stomach entirely empty, not a vestige of cabbage to be found. Extracted four drachms of gastric juice, mixed with a very little greyish white, chymous fluid.

Experiment 28.

Oct. 17. At 9 o'clock, A. M., he breakfasted on *stewed, salted pork, potatoes* and *bread.* Digested in three hours. Extracted gastric juice.

Experiment 29.

At 2 o'clock, 30 mins., P. M., same day, he dined on *boiled mutton, recently salted, squash, potatoes* and *bread.* Digested in three hours.

Some morbid spots begin to make their appearance on the mucous surface again; grumous blood exuding from several small points of the membrane; tongue slightly coated; countenance rather sallow; dull pain across the forehead, and through the eyes; appetite not impaired; at bed-time, put in through the aperture four drachms of *tinct. of aloes and myrrh,* diluted with water. This had the effect of correcting the morbid appearance of the stomach, and removed the pain in the head, &c.

Experiment 30.

Oct. 18. At 9 o'clock, 45 mins., A. M., he breakfasted on *boiled carrots,* and nothing else--full meal. 12 o'clock, M., examined stomach; considerable yellowish, pultaceous semi-fluid, remaining. 1 o'clock P. M., stomach empty.

Experiment 31.

At 7 o'clock, P. M., he ate three large *roasted potatoes,* with a little *salt*—nothing else. 9 o'clock 30 mins., stomach empty.

Experiment 32.

Oct. 19. At 9 o'clock A. M., he breakfasted on *broiled mutton* and *pancakes.* Digested in three hours and forty minutes.

Experiment 33.

At 2 o'clock, 15 mins., P. M., he dined on *stewed mutton* and *pancakes.* Digested in three and a half hours.

Experiment 34.

Oct. 20. At 9 o'clock, 45 mins., A. M., he breakfasted on one pint of *sago*, *boiled*, thick and rich, sweetened with *sugar*. 11 o'clock, 30 mins., stomach empty and clean.

There was no acrimony of the gastric contents, or smarting of the edges of the aperture, during the chymification of this meal, as is usual in most vegetable and farinaceous aliments; it seemed peculiarly grateful to the surface of the stomach; rendering the membrane soft, uniform and healthy.

Experiment 35.

At 12 o'clock, M., he ate four *eggs*, *roasted hard*, without any thing else. 3 o'clock P. M., stomach empty; no trace of the eggs to be seen.

Experiment 36.

At 4 o'clock 30 mins., P. M., he dined on *roasted duck* and *fried onions*. 8 o'clock, 30 mins., stomach not empty—distinct particles of food to be seen.—9 o'clock, stomach empty.

Experiment 37.

Oct. 21. At 9 o'clock, A. M., St. Martin breakfasted on one pint of *sago*, *boiled* and sweetened with *sugar*. 10 o'clock, 45 mins., stomach empty and clean; no vestige of the sago remaining; no acrimony of the gastric contents, or smarting of the edges of the aperture, during the chymification of this meal.

Experiment 38.

Oct. 22. At 12 o'clock, M., he ate four *fresh eggs*, *roasted hard*. 3 o'clock, P. M., stomach empty; no trace of the eggs to be seen.

At 4 o'clock, P. M., he dined on *roasted duck*, (domesticated,) dressed with *onions*. 8 o'clock, stomach empty.

Experiment 39.

Oct. 24. At 2 o'clock, P. M., he ate a pint of *soft custard*, and nothing else. 5 o'clock, 15 mins., stomach empty and clean.

At 6 o'clock, he ate three ounces of *strong old cheese*, and a piece of *bread.* 9 o'clock, 30 mins. stomach empty.

Experiment 40.

Oct. 26. At 9 o'clock, A. M., he breakfasted on *fricasseed chickens*, *bread* and *coffee.* 11 o'clock, 45 mins., stomach empty and clean.

At 12 o'clock, M., he dined on *roast chicken*, *bread* and *potatoes.* 4 o'clock, P. M., stomach empty.

Experiment 41.

Oct. 27. At 8 o'clock, A. M., he breakfasted on *broiled chicken*, *bread* and *coffee.* 11 o'clock, all digested, and stomach empty and clean.

At 12 o'clock, M., he dined on *chicken soup.* with *rice.* 3 o'clock, stomach empty.

At 5 o'clock, P. M., he ate a meal of *oyster soup* and *crackers.* 8 o'clock, 30 mins., stomach empty.

Experiment 42.

Oct. 26. 10 o'clock A. M., stomach empty, healthy and clean. I suspended through the aperture, into St. Martin's stomach, thirty grains precisely, of each of the following articles of diet, severally masticated and separately contained in small muslin bags, viz :—Fricasseed *breast of chicken; liver* and *gizzard* of do. ; *boiled, salted salmon ; boiled potato*, and *wheat bread* ; and he kept moderately exercising. At 3 o'clock, P. M., took out and accurately examined the several parcels. The breast of chicken was all digested and gone from the bag, to a mere atom, less than half a grain. The liver was almost as completely dissolved as the breast, half a grain only, remaining—of the bread, about the same ; less than a grain. The residuum of the gizzard, consisting principally of tendinous fascia, weighed seven and a half grains. The salmon, twelve grains. and the potato, six grains. The bags, containing these several articles, were attached to a string, at equal distances from each other, about an inch apart ; and

I allowed length enough for them to move freely through the stomach, and pass even to the pylorus. They were attached in the following order :—1st, the breast of chicken—2d, liver—3d, gizzard—4th bread—5th, salmon, and 6th, patato. When I withdrew them, they appeared to be retained quite forcibly at the pyloric end, requiring considerable force to start them at first, but after being drawn two or three inches, they came easily. The bags too, appeared to have been compressed, in proportion as they had been settled into the pyloric extremity, and were emptied in about the same proportion, with the exception of those containing the bread and potato, which, though above, had less remaining than that containing the gizzard. This, however, may be accounted for, from the more difficult solubility or digestibility of the tendinous parts of the gizzard. The bags seemed to have been as forcibly pressed, as if they had been firmly grasped in the hand. The four first on the string, (counting from the lower end upwards) more so, than the other two ; and the fourth more than the third. These circumstances coincide with the apparent contractions of a band, or circular muscle of the stomach, indicated by the motions of the glass tube, observed in former experiments. In comparing the length of the string, and situation of the bags, with the stem and bulb of the tube, it brought the fourth bag to that point in the stomach, where the contraction upon the bulb of the thermometer has invariably been observed to take place; the third bag just below, and the fifth and sixth, above it. The sensations expressed by St. Martin, on the extraction of these bags, were also indicative of the same facts. When I first commenced pulling the string, he complained of a sense of pain and distress at the pit of the stomach, and towards the pylorus, which increased while the bags were withdrawing, and particularly at this extremity, for the first three or four inches, till they had passed the band, into the splenic end.

The effects of this experiment, upon St. Martin's feelings and appearance, were very manifest, and afford interesting and important subjects of pathological consideration. He had not eaten or drunk any thing, that morning, and felt and looked in perfect health, when the bags were introduced; continued moderately exercising, and ate nothing but a small piece of dry bread, till they were taken out.

Soon after they were suspended in the stomach, he felt a sense of weight and distress at the scrobiculus cordis; slight vertigo and dimness of vision. These continued to increase and become quite severe, accompanied, at the latter part of the time, by slight pain in the forehead and through the eyes, and a sense of tightness or stiffness across the breast. His countenance had changed from a florid, healthy, to a sallow, sickly appearance, during the time of the experiment, and a soreness at the pit of the stomach continued after the extraction of the bags, for eight or ten hours, and had not entirely subsided the next morning.

Morbid action of the inner membranes was evident next day, with considerable erythema and aphthous appearance.

The first, second and third bags were covered with a thick mucous coat, tinged with yellow bile; the others had very little or none of this appearance.—This circumstance I conceive to have been owing to the irritation of the bag, at the pyloric extremity, inviting the bile from the duodenum to the stomach, in the latter part of this experiment. Hence the pathological indications which ensued. The same appearance and circumstances have before occurred during these experiments.

The following experiments on artificial digestion, were instituted with a view of ascertaining more particularly, the relative digestibility of many of the different kinds of aliment used in the foregoing gastric experiments, on natural chymification, and to test the correctness of the results. They are minutely detailed for the purpose of showing the manner, progress and operation of the gastric solvent, on the alimentary substances, subjected to its action. How far they may illustrate these subjects, the reader will judge for himself.

The gastric juice was taken out of the stomach in different states of purity and put into vials; when food was submitted to its action, it was placed in a temperature between 96° and 100°, Fahrenheit, and kept either in the axilla, or on a sand bath, and frequently, though not constantly, agitated.

The discrepance of results in some similar experiments will generally be found to arise from the variable degrees of purity of the gastric juice, or different circumstances of the experiments.

Experiment 43.

Sept. 18. At 8 o'clock, 45 mins., A. M., I put one drachm of *fresh salmon trout, fried*, and masticated, and one drachm of *wheat bread*, into two ounces of gastric juice, taken from the stomach yesterday and this morning. The juice was not perfectly clear, but contained some viscid mucus. Placed them in the axilla and kept moving. 10 o'clock, 15 mins., residuum of aliment taken out, filtered and pressed as dry as when put in, weighed one drachm and five grains. The menstruum, after filtering, was white and opaque, about the colour and consistence of rich gruel. Mixed the residuum and fluid together again and placed the vial on the sand bath, and kept it constantly agitated for *one hour*. Taken out, filtered and dried as before, the undissolved residuum now weighed just *thirty grains*. The fluids had become thicker, and richer in colour and consistence. Put them together again into the vial, and continued them on bath and in axilla, *another hour*, though not so constantly agitated, as during the last hour. The residuum, treated in the same manner as before, now weighed *twenty four grains.* Mixed together and continued in axilla *two hours* more, the residuum weighed *twelve grains.* After continuing *three hours* longer in the axilla, the undissolved portions of aliment, consisting principally of particles of fish skin; weighed *four grains*, which became gradually diminished during its continuance *an hour* longer in the axilla.

The menstruum at this time, was of a rich gruelly colour and consistence, slightly tinged with a reddish cast, or colour of the fish. Set this aside for thirty eight or nine hours.

Sept 20. 9 o'clock, A. M., food almost entirely reduced to chyme, of a rich, lightish coloured, gruelly appearance; some few particles of the skin of the fish remaining undissolved, with some small, apparently foreign and indigestible substances, which were probably adventitiously mixed with the food.

To observe the effect produced on this chyme, by the addition of bile, and having very opportunely obtained some, from the human stomach, by the operation of an emetic, I added *one drachm* of this pure, albuminous, orange coloured bile, to six drachms of the chyme. The first apparent change, was in the colour, which partook of the bile; then a slight effervescence was perceived, and very fine coagulæ were formed. The fluid became richer in appearance, and less opaque. The foreign or indigestible particles, were more perceptible, and small, bright particles, resembling very minute scales, or skin of fish, were also quite plain to be seen.

I now divided this into two equal parts; to one of which, I added half a drachm of dilute muriatic acid, and set it by to subside. Examined at 10 o'clock, the 21st. The vial containing the mixture of chyme, bile and muriatic acid, exhibited the following appearance: It had a thick, dense sediment, of a yellowish green colour, which occupied about one quarter, of the space. The fluid above, was of the colour of whey, and about the consistence. The vial containing the mixture of chyme and bile only, showed the following appearance: The sediment was not so dense, and its colour, as well as the supernatant liquid, was rather more yellow. Standing at rest a few days, the sediment, at the bottoms of both vials, became more compact; that in the muriatic mixture, more so than the other, and was of a deeper green colour; the fluid continued of a rich, whey colour and consistence, and a very thin pellicle, or small whitish flocculi, rose on the top, or adhered to the sides of the vial.

Experiment 44.

Sept. 20. At 1 o'clock, 15 mins., P. M., I put one drachm of *boiled*, *green corn* and *beans*, into twelve drachms of gastric juice, and kept the vial in the axilla, or on the bath, as usual, frequently agitating it, till 7 o'clock, P. M. The residuum, at this time, taken out, weighed *twenty eight* grains, consisting wholly of the hulls or cuticular parts of the broken kernels, and one entire bean and a kernel of corn; the first of which weighed thirteen, and the other eleven grains, leaving four grains of the skins of the broken, dissolved grain. The two entire kernels, (the bean and the corn) were designedly put in whole, to test the effect of the gastric juice upon them, in the entire state. The other portion of the grain was mashed soft before put in. The pulpous portion of the broken kernels was all dissolved, and appeared completely chymified. The fluid was nearly as white as milk, and of the consistence of clear rich gruel.

The gastric juice used in this experiment, was considerably vitiated when taken from the stomach, some thirty six or forty eight hours previously, and was quite fœtid when used. This fœtor was, in a great measure, corrected after chymification of the food had commenced; the sharp, acid flavour, so peculiar to forming chyme, was increased.

Experiment 45.

Sept. 21. At 8 o'clock, 15 mins., A. M., I put

thirty grains of *fresh beef steak* and thirty grains of *fresh beef's liver*, (broiled, and masticated) contained loosely in separate muslin bags, into one ounce of fresh gastric juice, and kept them in axilla. At 9 o'-clock, 45 mins., the two parcels of aliment, taken out and pressed as dry as when put in, weighed as follows: The *steak*, seventeen grains; the *liver*, eleven grains. Put into the vial again, and continued in the axilla, till 1 o'clock, P. M. The *steak* weighed fourteen, and the *liver* eight grains. Put into the vial again and continued in axilla for four hours; no further change was effected. They both weighed the same as at last examination. The solvent action having ceased, I added one ounce more of gastric juice, and continued in axilla, two hours and thirty minutes. The *beef* weighed five grains, and the *liver*, four; the residue of the liver consisted mostly, of membranous particles, like sections of the hepatic blood vessels, of which I conceived them to be portions.

I now mixed them both together, in one bag, and continued them in axilla, three hours, when the whole were completely dissolved and chymified, and the bag empty; with scarce a trace of aliment left on the inside. The fluid was of a greyish white, gruelly appearance. A brownish sediment was deposited on standing.

Experiment 46.

Sept. 22. At 12 o'clock, 30 mins., I put thirty grains of *new cheese*, (masticated) into three drachms of gastric juice, and placed it in the axilla, eight hours and thirty minutes, when five grains of the cheese remained undissolved, or rather unchymified, as the residuum was in nearly a liquid form, consisting, principally, of oil, combined with a soft caseous substance, floating on the surface of a rich, milky fluid. A little very fine, white, compact sediment, at the bottom of the vial. At this time, it had acquired a strong acid, or peculiar acrid taste, and emitted a strong, caseous smell, even stronger than the cheese itself presented, when put in.

Experiment 45.

At 12 o'clock, M., I put one drachm of *sago*, boiled so as to leave some of the grains whole and entire, but soft and gelatinous, into three drachms of gastric juice and kept it in the axilla. When first mixed, they were so much alike, that they could only be distinguished from each other by the globular forms of the grain. But by these, however, the gastric juice could distinctly be perceived to dissolve the grains of sago, till they had all disappeared.

The fluid had now become more opaline and whitish, and in two hours and twenty minutes, no trace of the sago could be discerned. At this time the fluid had become more opaque and milky. No sediment was deposited on standing for twenty four hours. A slight acid was perceptible.

Experiment 46.

At 1 o'clock, P. M., I took three vials, the first containing two drachms of gastric juice; the second, two drachms of common vinegar; and the third, two drachms of simple water. Into each of these, I put ten grains of *raw albumen* of a fresh egg.—When first put together, they presented the following appearances: The albumen put into the gastric juice, at a temperature of about 76°, produced loose coagulæ in a few minutes, generally diffused through the juice, but soon collected into a more compact mass, and subsided towards the bottom of the vial. That put into the vinegar, produced similar coagulæ and loose mass, and fell down. That in the vial of water produced loose, light coloured flocculi, equally suspended through the water, but not inclining to collect together, like the other two.

These three parcels, kept in the axilla, and agitated for two hours, presented the following appearances: The coagulæ in the gastric juice, was half dissolved, and the menstruum of a milky appearance.

Those in the vinegar and water, remained the same, and their fluids unaltered. In five hours, that in the gastric juice was entirely dissolved, and the fluid more opaque and white; the other two remained of the same appearance as at last examination; the coagulæ in the vinegar, taken out, weighed *nine* grains—that in the water was too loose and frothy to be collected and weighed.

Experiment 47.

Sept. 25. At 7 o'clock, A. M., I put twenty grains of light *sponge cake* into three drachms of gastric juice, and kept it in axilla. It was all dissolved and chymified, in seven hours. The fluid was rich, yellowish white, or cream colour, and of the consistence of gruel, with a little loose, brown sediment at the bottom of the vial, after standing.

Experiment 48.

At 9 o'clock, A. M., I put two *purple fox grapes*, one skinned and the other entire, into six drachms of gastric juice, and kept them in axilla, six hours, with very little alteration in their appearance; the skinned grape, weighing, when first put in, thirty four grains, weighed now, thirty grains, retaining its shape and texture. The whole grape was not affected in the least, either in shape, colour or texture. It weighed fifty four grains when put in, and the same now. Continued in axilla, twelve hours, they remained unaltered, and weighed exactly the same as at last examination. Added one ounce of fresh gastric juice, and continued them in axilla, twenty four hours, unaltered. The texture of the skinned grape, was as firm and hard as when first put in; and the fluid was unchanged in its appearance, except a slight fœtor, perceptible at the end of three or four days.

This, I think, is a fair specimen of the indigestible nature of this kind of fruit.

Experiment 49.

Sept. 26. At 10 o'clock, A. M., I put thirty grains of *ripe, mellow peach*, and thirty grains of *ripe, hard apple*, into one ounce of gastric juice, and kept them in axilla, till 8 o'clock, P. M. At this time the residuum of the peach, weighed eighteen grains—the apple, twenty four grains. They were neither of them mashed or masticated, but cut into small, square pieces, strung on a string, and suspended into the juice in a vial.

At 10 o'clock, A. M., of the 27th, after having been continued in axilla, six hours longer, the peach weighed ten grains and the apple the same as at last examination, twenty four grains. The peach had now become soft and pulpous, and fallen from the string. Eight hours longer continuance in axilla, completed the digestion of the peach; but the apple remained nearly the same.

Experiment 50.

Sept. 27. At 2 o'olock, P. M., I put one drachm of *albumen* of egg into four drachms of gastric juice, fresh from the healthy stomach. At first, the albumen fell to the bottom of the vial; but on being agitated, it was diffused through the juice, and in a few minutes, loose coagulæ formed, and remained suspended near the bottom of the fluid. Raised the temperature to 100°, and placed the vial in the axilla.

At the same time, I put one drachm of *albumen* into four drachms of simple water, at the same temperature, and placed it with the other in the axilla. When first put together, the albumen was diffused, in loose, light flocculi, through the water, not coagulating and collecting like that in the gastric juice, and subsiding to the bottom, but adhered to the sides of the vial, or rose to the surface.

When both vials were smartly agitated, a white, frothy mass, formed on the top of the water, filling the two ounce vial in which it was contained. The vial of albumen and gastric juice exhibited the coagulæ, broken into small particles, falling towards the bottom again. Kept in the axilla and frequently agitated, for one and a half hour, the gastric mixture had become semi-opaque and the coagulæ considerably diminished in quantity. The aqueous mixture remained unchanged; the frothy portion on top, and the fluid, perfectly limpid and clear, below. No appearance of the albumen in any shape, could be seen, except the floating froth. Indeed, the albumen seemed to have clarified the water, and rendered it clearer than at first. At 6 o'clock, P. M., the albumen in the gastric juice was completely dissolved; the fluid was white and milky, with a little very fine, dirty white precipitate falling to the bottom, on standing at rest. That in the water was strikingly different in appearance. The agitation had beaten up the albumen completely into beautiful white froth, and it lay like a snow ball or bunch of clear, raw cotton, on the surface of the water, now transparent as crystal, without the least particle of sediment to be seen.

At 7 o'clock, I added two drachms of gastric juice to the vial containing the water and albumen, and continued it in axilla. In two hours, the solvent effect of the juice, upon the frothy mass, was very evident. The white froth upon the top, was almost entirely diminished and gone. Neither could agitation re-produce it as at first; small white coagulæ, like those seen in the other vials, were now distinctly visible; the fluid had become opaque and whitish, like the other, and a little fine sediment settled to the bottom, on standing. Continued in the axilla, two hours longer, it resembled almost exactly, that in the other vial, in every particular.

Experiment 51.

At 2 o'clock, P. M., I put one drachm of *yolk of egg* into four drachms of gastric juice, and another drachm into four drachms of simple *water*, and kept them, as usual, in the axilla; no difference at first could be perceived between the gastric juice and aqueous mixtures ; each exhibited a yellow mixture, like the egg, simply beat up with any white or watery menstruum. Six hours continuance of this treatment, produced little difference in the appearance of the two, and effected a slight modification in the gastric mixture only ; this seems to have been converted into a very fine coagulæ, of a rich cream colour and consistence, and of a paler yellow than the other. In twelve hours more, a striking difference was manifest—that in the water remained the same as when first put together—a dull, yellow coloured sediment, in the proportion of about one fifth of the space occupied by the whole, had subsided to the bottom of a thin fluid, of the same colour, and now emitted a fœtid odour. That in the gastric juice had become more cream-like and lighter coloured, separating, on standing, into three distinct portions—a loose, coagulated, yellow mass, rose to the top, occupying more than half the upper space—a clear, whey-coloured fluid below, with a dirty, yellow sediment at the bottom, in about the proportion of one twelfth of the whole ; not the least fœtor was perceptible.

Experiment 52.

At 1 o'clock, 30 mins., P. M., I mixed one drachm of *Olive oil* with three drachms of gastric juice, and kept, frequently agitated in axilla, for eight hours.—When first put together and shaken, the mixture resembled water and oil, precisely ; after continuing in the axilla, four or five hours, the oil had perceptibly diminished and chyme began to be formed, rendering the juice opaque and milky. At 10 o'clock, P. M., the oil was about one sixth diminished, the menstruum nearly the colour and consistence of milk.

Sept. 30. 8 o'clock, A. M., continued in the same manner, in the axilla for twelve hours, the oil was proportionally diminished, and the opacity and milkiness, gradually increased.

Oct. 1. At 8 o'clock, A. M., I added one drachm of gastric juice—not clear, but considerably vitiated. Continued in axilla fourteen hours. Similar proportional decrease of the oil, and change of the colour of the fluid, were produced, and a slight fœtor was perceptible. This last circumstance, no doubt was attributable to the vitiated juice added.

Oct. 2. 10 o'clock, A. M., added three drachms of pure gastric juice, and continued in axilla, *ten* hours. This addition corrected the fœtor in a great measure. The stratum of oil was not much diminished in bulk, but considerably changed in appearance, having become quite white and frothy, exhibiting myriads of minute globules; and the colour and consistence of the fluid, were more rich and milky.

On the 3d at 10 o'clock, A. M., I divided the contents of the vial into two equal parts, and put them into two separate vials. To No. 1, I added two drachms of pure gastric juice; and to No. 2, two drachms of fresh extracted gastric juice, containing a large proportion of yellowish green bile, and con tinued, as usual, in axilla. The following changes were produced: The portion in No. 2 vial, which had received the yellow gastric juice, at first partook of the yellow colour of the juice added, generally diffused through the whole mass—a separation then took place; the bile seemed principally to unite with the oil, breaking it down and reducing it to very minute and almost imperceptible globules; and after remaining in the axilla, ten hours, and then standing at rest a few minutes, the under surface of the supernatant stratum of oil exhibited a milky or creamy appearance, and small, white flocculi, resembling coagulated milk or albumen; these soon became dissolved, and increased the richness of the fluid below—No sediment to be seen. The portion in No. 2 vial, to which the clear gastric juice was added, at the end of ten hours, had undergone some change. The pellicle of oil on the surface, was reduced to minute globules, of a whitish colour. The same appearance of white flocculi, or coagulæ, were exhibited upon the under surface of the supernatant stratum of oil, as in the other, but not so abundant, and the fluid was not so rich in colour and consistence.

Oct. 4. At 9 o'clock, A. M., I added two drachms more of each kind of juice, to their respective parcels, and continued them as usual, in axilla, for eleven hours. The difference between the two parcels, was now considerably increased. The fluid in No. 2 vial, was of a rich cream colour and consistence; the supernatant stratum of oil was converted into a light yellowish mass, resembling a mixture of gelatine and coagulæ; few of the globules of the oil could be distinguished; yellow flocculi adhered to the sides of the vial, above the fluid, after being agitated. When suffered to stand at rest a short time, loose yellow flocculi rose on the surface, occupying more than twice the space of the oil, before the last addition of gastric juice—no sediment subsided.

The parcel in No. 1 vial, had regularly progressed in chymification, in ratio proportional to the juice added; the supernatant, oily stratum was diminished, in thickness, nearly one third, since the last addition of gastric juice; had changed from its oily appearance, into a white, semi-gelatinous mass, intermingled with milk white flocculi; the fluid of the same milky appearance; a little white sediment at the bottom.

Oct. 5. At 10 o'clock, A. M., I added six drachms pure gastric juice, and six drachms of fresh extracted juice, containing about the same proportion of yellow bile as the other, to their respective vials, and put them on the bath, and kept them continually agitated for five hours. The effect was palpable and plain. The supernatant stratum, in No. 2 vial, was now completely broken down, and not a globule remained; a thin, yellow pellicle, or loose flocculi, rose upon the surface, on standing, and the fluid was of a rich, cream colour and consistence, slightly tinged with bile—no sediment perceptible.

The contents of No. 1 vial, had undergone considerable change; the oily pellicle on the surface, was diminished but little in volume, but changed in appearance; had become converted into a white semi-gelatinous, or rather saponaceous consistence, and the milky richness of the fluid was increased.

This experiment is minutely and accurately detailed, with a view to demonstrate the slow, but certain digestibility of oils, and the manner they are acted upon by the gastric juice. It may be tedious, from its prolixity, but I considered its communication might be of some importance and usefulness to physiological science, the interests of which have been of paramount consideration with me, in all these experiments.

It very clearly appears, by this experiment alone, that *bile* accelerates the solution of oil, by the gastric juice; and I have no doubt, it facilitates the chymification of all fatty and oily aliments; and is required, and necessarily called into the stomach *only* for that purpose. This has been frequently indicated in the course of these experiments, by the effect which it has produced on fatty or oily aliments, when adventitiously mixed with the gastric juice.

Experiment 53.

Sept. 29. At 1 o'clock, P. M., I mixed one drachm of sweet *cream*, with three drachms of clear gastric juice, and placed them in the axilla. When first put together, the juice fell to the bottom of the vial, and remained distinctly separate from the cream, till agitated, when they united, but exhibited no other immediate change of appearance. When the temperature was raised to above 80°, the whole gradually formed into very fine, creamy coagulæ. Continued in axilla twelve hours, this coagulated mass was more than half diminished, and rising to the top of an opaque white, whey-coloured liquid. Small globules of oil were now seen on the upper surface of the supernatant coagulæ—No sediment.

Oct. 1. 10 o'clock, A. M., I added one drachm of clear gastric juice, and continued in axilla, ten hours, when the creamy coagulæ were still more diminished; the globules of oil on the surface, increased, and the liquor below, resembled clear, rich gruel, occupying about one sixth of the space of the whole.

Oct. 2. 12 o'clock M., I added another drachm of gastric juice, and continued it in axilla, eight hours. The creamy coagulæ were now reduced to about one fourth, and more loose and white than at first. The globules of oil were now much increased, and formed a complete pellicle over the whole upper suface, nearly resembling soft butter, and emitted a slight rancid flavour. The richness of the chymous liquid below was proportionally increased. No sediment.

Oct. 3. 12 o'clock, M., I divided the contents of the vial into two equal parts, and put them into two separate vials. To No. 1, I added two drachms of pure gastric juice; and to No. 2, two drachms of fresh extracted gastric juice, strongly tinged with yellowish green bile, and kept them in axilla nine hours. The changes effected, after this addition, were strikingly evident, and different in the two parcels. That in No. 2, to which was added the yellowish green juice, exhibited a perfectly homogeneous, rich, gruel-like liquid, slightly tinged with the bile; the creamy coagulæ were all dissolved, and not a globule of the oil to be seen; all appeared chymified —a little dirty white sediment fell to the bottom.

The creamy coagulæ of No. 1 vial, were not completely dissolved, but reduced to a thin, loose layer, and the oily pellicle was scarcely perceptible; the globules extremely minute and whitish, and of a saponaceous consistence. The fluid below, was of a light coloured, rich, gruelly appearance. No sediment deposited. To complete the chymification of the contents of No. 1, I added two drachms more, clear gastric juice, and continued it in axilla, twelve hours longer: at the end of this time, the coagulæ were reduced to a very thin layer; the oily pellicle entirely dissolved, and the liquid of a rich gruelly colour and consistence. No sediment subsided on standing.

Experiment 54.

Oct. 1. Mixed four drachms of *sweet, skimmed milk* with four drachms of gastric juice, and kept in axilla. The juice fell to the bottom, when first put together, as with the cream; but when shaken, and raised to 90° or 100° temperature, formed into loose and coarser coagulæ, than the cream, which were diffused and suspended through the milky fluid. Continued in axilla eight hours, the coagulæ were more collected, firmer and more than half diminished.—The fluid of a light whey, or thin gruel colour and consistence, with a few loose, white flocculi, and a creamy pellicle on the top.

Oct. 2. Continued in axilla eight hours more, the coagulæ were almost completely dissolved; fluid the colour of rich strained gruel; a few light flocculi on the surface, but no creamy pellicle; a little coarse sediment, or loose, white coagulæ at the bottom.

Experiment 55.

Oct. 3. Put fifteen drops of gastric juice into three drachms of *sweet milk*, at the temperature of 65°; a slight appearance of very fine coagulæ, was first exhibited, but not so as to become distinctly separated, till after the temperature was raised to 85° or 90°, when the whole mass gradually formed into a tremulous, jelly-like curd, which, after cooling, and standing at rest a few hours, separated into two about equal parts; a soft, caseous substance, and a thin, light coloured whey.

Experiment 56.

Oct. 3. Put two drachms of the *soft, caseous substance*, formed in the above experiment, (55) into one ounce of gastric juice, and placed in axilla, six hours; at the end of this time, the curd, or caseous substance, was nearly all dissolved; the menstruum of a white gruel-like appearance, with a thin pellicle of loose, white coagulæ on the surface. In four hours more, it was all dissolved; the fluid richer, and perceptibly acid.

Experiment 57.

Oct. 13. 9 o'clock, A. M. Into one ounce of gastric juice, I put one and a half drachms of the *medulla spinalis* of an ox, enveloped in its neurilema, boiled, and placed it on the sand bath, or in axilla, six hours. At 3 o'clock, P. M., examined—the medulla had fallen out of its envelope, and when taken out and separated from the fluid, by the filter, weighed fifteen grains; the neurilema, at the same time, weighed eighteen grains. Put these remaining portions into two drachms fresh gastric juice, and continued in axilla six hours. At 9 o'clock, P. M., the remainder of the medullary portion, weighed eight grains, and the neurilema, nine grains. Continued in axilla, three hours longer, the medullary part weighed three grains, and the neurilema, four grains. The menstruum was now a rich, milk white liquid, of nearly the consistence of cream. A loose, light sediment fell to the bottom, on standing; the fluid retained its rich, milky whiteness and creamy consistence.

Experiment 58.

Oct. 14. 9 o'clock, A. M., put half a drachm of *medullary substance*, the brain of an ox, boiled, into four drachms of gastric juice, and kept it on the bath, frequently agitated, six hours, when it was all dissolved, and had produced a rich milky fluid, with a loose, light sediment.

Experiment 59.

Oct. 15. Put twelve grains of solid *beef bone*, broken into small pieces, with the periosteum attached to one side, into one ounce of fresh gastric juice, and kept in axilla, twelve hours. At this time the periosteum was nearly dissolved; weight of the bone, ten grains. Added six drachms of gastric juice, considerably vitiated, and continued in axilla nine hours, and the bone weighed nine grains. The menstruum was now a whitish opaque fluid, about the consistence of clear, thin gruel, with a little light brown sediment, settling to the bottom, on standing. Added one ounce more gastric juice, and continued it in axilla, twelve hours. The weight of the bone, at the end of this time, was six grains. The opacity and richness of the fluid increased; smell, slightly fœtid. Discontinued the experiment.

The result of this, confirms the correctness of some former observations, in similar experiments, and sufficiently demonstrate the solubility of solid bone, in the gastric juice of the human stomach.

Experiment 60.

Oct. 17. 1 o'clock, P. M., I put twenty grains of boiled *mutton suet*, cold, and divided into small pieces, into six drachms of gastric juice, tinged with bile, and kept it in axilla, *seven hours*. The undissolved residuum, separated by the filter, now weighed ten grains; and the fluid was as white as milk, and about the consistence of thick gruel; there was no appearance of any oily particles; it seemed to have been coagulated, and converted into chyme, like milk or albumen; the chymous part very much resembled that formed from medullary substance.—Continued on axilla, three hours longer, it was all dissolved, and the richness of the fluid considerably increased.

Experiment 61.

Oct, 25. 2 o'clock, P. M., put one drachm *custard*, into one ounce of gastric juice, fresh from the stomach, and placed it in axilla. 8 o'clock, 30 mins., all dissolved and chymified; fluid, as usual, from such aliment, of colour and consistence of rich gruel.

Experiment 62.

Nov. 1, 1833. To one ounce of gastric juice, *taken from the stomach in Dec.* 1832, (and which was as pure as when first extracted,) I added thirty grains of *lean, boiled mutton*, masticated. Kept in axilla, six hours, it dissolved sixteen grains. The fluid exhibited the usual appearance of chyme.

TABLE,

Showing the mean time of digestion of the different Articles of Diet, naturally, in the Stomach, and artificially, in Vials, on a bath.

The proportion of gastric juice to aliment, in artificial digestion, was *generally* calculated at one ounce of the former to one drachm of the latter, the bath being kept as near as practicable at the natural temperature, 100° Fahrenheit, with frequent agitation.

Articles of Diet.	*Mean time of chymification*			
	In Stomach.		In Vials.	
	prep.	h. m.	prep.	h. m.
Rice, -	boiled	1 00		
Sago, -	do.	1 45	boiled	3 15
Tapioca, -	do.	2 00	do.	3 20
Barley, -	do.	2 00		
Milk, -	do.	2 00	do.	4 15
Do. -	raw	2 15	raw	4 45
Gelatine. -	boiled	2 30	boiled	4 45
Pig's feet, soused,	do.	1 00		
Tripe, do.	do.	1 00		
Brains, animal,	do.	1 45	do.	4 30
Venison, steak,	broiled	1 35		
Spinal marrow, animal,	boiled	2 40	do.	5 25
Turkey, domesticated,	roasted	2 30		
Do. do.	boiled	2 25		
Do. wild,	roasted	2 18		
Goose, do.	do.	2 30		
Pig, sucking -	do.	2 30		
Liver, beef's, fresh,	broiled	2 00	cut fine	6 30
Lamb, fresh,	do.	2 30		
Chicken, full grown,	fricas'd	2 45		
Eggs, fresh,	h'rd bld	3 30	h'rd bld	8 00
Do. do.	soft bld	3 00	soft bld	6 30
Do. do.	fried	3 30		
Do. do.	roasted	2 15		
Do. do.	raw	2 00	raw	4 15
Do. whipped.	do.	1 30	whipped	4 00
Custard, -	baked	2 45	baked	6 30
Codfish, cured dry,	boiled	2 00	boiled	5 00

Articles of Diet,	*Mean time of chymification.*			
	In Stomach.		In Vials.	
	prep.	h. m.	prep.	h m.
Trout, salmon, fresh,	boiled	1 30	boiled	3 30
Do. do.	fried	1 30		
Bass, striped, do.	broiled	3 00		
Flounder, do.	fried	3 30		
Catfiish, do.	do.	3 30		
Salmon, salted,	boiled	4 00	do.	7 45
Oysters, fresh,	raw	2 55	raw,entir	7 30
Do. do.	roasted	3 15		
Do. do.	stewed	3 30	stewed	8 25
Beef, fresh, lean, rare,	roasted	3 00	roasted	
Do. do. dry	do.	3 30	do.	7 45
Do. steak,	broiled	3 00	mastic'd.	8 15
Do. do.	do.		cut fine	8 00
Do. do.	raw		do.	8 15
Do. with salt only,	boiled	3 36		9 30
Do. with must'd. &c.	do.	3 10		
Do. fresh, lean,	do.		mastic'd.	
Do. -	do.		entire p.	9 00
Do. -	fried	4 00		12 30
Do. old, hard salted,	boiled	4 15		
Pork, steak, -	broiled	3 15		
Pork, fat and lean,	roasted	5 15		
Do. recently salted,	boiled	4 30	mastic'd.	6 30
Do. do.	fried	4 15		
Do. do.	broiled	3 15		
Do. do.	raw	3 00	raw	8 30
Do. do.	stewed	3 00		
Mutton, fresh,	roasted	3 15		
Do. do.	broiled	3 00	mastic'd.	6 45
Do. do.	do.		unmas'd.	8 30
Do. do.	boiled	3 00		
Veal, fresh, -	broiled	4 00		
Do. do. -	fried	4 30		
Fowls, domestic,	boiled	4 00	mastic'd.	6 30
Do. do.	roasted	4 00		
Ducks, domesticated,	do.	4 00		
Do. wild,	do.	4 30		
Suet, beef, fresh,	boiled	5 30	entire p.	12 00

Articles of Diet.	*Mean time of chymification.*			
	In Stomach.		In Vials.	
	prep.	h. m.	prep.	h. m.
Suet, mutton,	boiled	4 30	divided	10 00
Butter, -	melted	3 30		
Cream, -			raw	25 30
Cheese, old, strong,	raw	3 30	mastic'd.	7 15
Do. do.			entire p.	18 00
Do. new, mild,			divided	8 30
Oil, Olive, -			raw	60 00
Soup, beef, veg. & br'd.	boiled	4 00		
Do. marrow bones,	do.	4 15		
Do. bean,	do.	3 00		
Do. barley,	do.	1 30		
Do. mutton,	do.	3 30		
Green corn & beans,	do.	3 45		
Chicken soup,	do.	3 00		
Oyster soup,	do.	3 30		
Hash, meat & veg.	warmed	2 30		
Sausage, fresh,	broiled	3 20		
Heart, animal,	fried	4 00	entire p.	13 30
Tendon, -	boiled	5 30	mastic'd.	12 45
Do. - -			entire p.	24 00
Cartilage, -	do.	4 15	mastic'd.	10 00
Do. - -			divided	12 00
Aponeurosis,	do.	3 00	boiled	6 30
Bone, beef's, solid,			entire p.	80 00
Do. hog's, do.			do.	80 00
Beans, pod,	do.	2 30		
Bread, wheat, fresh,	baked	3 30	mastic'd.	4 30
Do. corn, ·	do.	3 15		
Cake, do. -	do.	3 00		
Do. sponge,	do.	2 30	broken	6 15
Dumpling, apple,	boiled	3 00		
Apples, sour, hard,	raw	2 50	entire ps.	18 00
Do. do. mellow,	do.	2 00	mastic'd.	8 30
Do. sweet, do.	do.	1 30	do.	6 45
Parsnips, -	boiled	2 30	mashed	6 45
Do. -	do.		entire p.	13 15
Do. -	raw		do.	18 00
Carrot, orange,	boiled	3 15	mashed	6 15

Articles of Diet.	*Mean time of chymification.*			
	In Stomach		In Vials	
	prep.	h. m	prep	h. m
Carrot, orange,			entire p.	12 30
Do. -			raw, do.	17 15
Beets, -	boiled	3 45		
Turnips, flat,	do.	3 30		
Potatoes, Irish,	do.	3 30	mashed	8 30
Do. do.			entire p.	14 00
Do. do.	roasted	2 30		
Do. do.	baked	2 30		
Cabbage, head,	raw	2 30	mastic'd	12 30
Do. with vinegar,	do.	2 00	shaved	10 15
Do. -	boiled	4 30	boiled	20 00
Peach, mellow,			cut small	10 00
Do. do.			mashed	6 00

The foregoing table is formed from all the experiments made upon St. Martin, since 1825, taking the average from such as were generally performed under the naturally healthy condition of the stomach, and ordinary exercise.

The mean times of artificial chymification, have been taken from such experiments as were generally made with the pure gastric juice, or such as was too slightly vitiated, to impair its solvent effect, in any essential degree.

They exhibit the average, as near as practicable, for the digestion of one drachm of alimentary matter, in one ounce of gastric juice, or in about that proportion, counting the time actually kept on the bath, or in the axilla.

Exceptions, however, must be made for the bone, oil, cream, and one or two other articles, which chymify much slower and more difficultly, than the less concentrated aliments.

Several experiments have been omitted, especially when they were of the same kinds, and produced similar results.

Showing the temperature of the interior of the Stomach, in different conditions, taken in different seasons of the year, and at various times of the day, from 5 o'clock in the morning till 12 o'clock at night.

Date.	Wind and	Weather	Th.	Temp're & condit'n of stomach — Empty		Dur'g chym'n		
1829.				repose	ex'rcis	repose	ex'rcis	
Dec 6	S	Cl'dy and damp	63	98°				
7	N W	do do	27	98				
8	S W	Clear and dry	13	99				
9	W	Clear	10	99				
1830.								
Jan 24	N W	do and cold	0.3	100				
25	S W	do	2	100		100		
Mar 17	S W	Rainy	38	99				
18	N W	Clear	6	100			102	
19				98				
1832								
Dec 4	N W	Snowing	35		101			
5			30	100			101 1 2	
6			38	100				
7			28	99		100		Stomach morbid.
8		Cl'dy and damp	46	99		99		do do
13				100				
14				100				Stomach morbid.
15				100				do do
22				100		100		
23				100	101			Stomach morbid.
25	E	Variable	31	100		100	101	do do
26	N E	Cl'dy and damp	38	99 1-2	101	99 1.2	101	
27	E	Foul and damp	38	99 1-2		100		
"	S	Clear	62	100		100		
28	N	do	34	100				
29	N W	do	34	100		100		
30	"	do	26	100				
31	S	Cl'dy and damp	30	100 1-2				Stomach morbid.
1833.								
Jan 1	S	Rainy	50	100				
3		Clear	33		101 1.2			
7	N E	Cl'dy and damp	43	100				
11	S W	Clear	15	100				
13	C'lm	Cloudy and dry	12	100	101	100	100 1 2	Stomach morbid.
14	N W	Clear	28	100			101 1 2	
15	N E	Cloudy and dry	35	100	101			
17	N W	Clear and dry	19	100		100	102	stomach morbid.
23	N E	Rainy	39	100 1 2			101 3.4	
24	N	Cl'dy and damp	39	100 1 2	101 1·4			
"	N E	Rainy		99 1 2				after sleeping.
25	S		36	99				before rising.
"			38	100 1.2			102	
26	N W	Clear	36	100 1 2		100 3.4	101	99 1 2 aft. sl'p'g.
27	C'lm	Cloudy	32	99 1 2			101 1.4	99 1 2 bef. rising.
28	S W	Clear	35	101 *			101 1.2	
"	S W	do	46	101 1 2		101 1.2		
"				101 1.2		101 1.2		
29	N E	Clear	28	100 3 4	101 1.2		102	100 before rising.
30	N E	Cl'dy and damp	39	99 1·2	101 1.2	101 1.4	102	99 1 2 bef. rising
31	N E	Rainy	45	101 1 4	101 1.2	101 1.4		100 do do
Feb 1	N W	Clear	28	101			102	100 do do
Mar 26		do		100 1·2			101	
July 9	W	Cldy and damp		100				before rising
10	W	Clear	63	100	101			
11	N E	Cloudy	65	100	101			

1833.				Temperature and condit'n of stomach.			
				Empty		*During Chym'n*	
	Wind & Weather.		*Th*	*Repose*	*Exercise.*	*Repos*	*Exercise*
July 12	W	Clear	70	100 1-2	101 3.4		
13	C'm	Clear	69	100 3-4	101 3.4		
14	S	Variable	75	100	102		
15	W	Clear	74	100	102		
	W	Clear	74	100 1.2			101 3.4
16	W	Cloudy	73	101	101 1-2		
28	N W	Clear	66	101			
Oct. 10	W	Fair	61	100	101 1-2		101 1-2
	S	Rainy	61		101 3.4	102	103
11	N W	Fair	32	100	102		102
11				101 3.4		101 3-4	
12	S	Cloudy	36	101		101	
13	N E	Rainy		101		101	102

Abstract of Temperature of the Stomach.

When empty, and in repose, highest, 100 3-4 }
do. do. do. lowest, 98, } Mean, 100 1-3°

do. do. exercising, highest, 102, }
do. do. do. lowest, 100, } Mean, 101 1-2°.

Full, or during chym'n, in rep. highest, 102, }
do. do. do. lowest, 99, } Mean, 100 1-5.°

do. during chym'n, in exerc., highest, 103, }
do. do. do. lowest, 100 1 2, } Mean, 101 1-2°.

* At this, and the subsequent examinations, the bulb of the thermometer was placed three or four inches nearer the pylorus than before, and exhibited an increase of temperature, indicating a difference of three fourths of a degree, between the splenic and pyloric extremities.

In all the observations previously to the 4th of December, 1832, the examinations were made with a Mercurial Thermometer, (Fahrenheit's) and north of latitude 43°. Subsequently, and to March, 1833, the examinations were made at Washington, D. C. in Lat. 38° 53', with the Spirit Thermometer, from Pool's Barometer, which varied half a degree from the mercurial one. From July 9, to November, 1833, I used Pool's Glass Chemical Spirit Thermometer, (Fahrenheit's scale) at Plattsburgh, N. Y., in latitude about 44° 40', N.

INFERENCES.

From the foregoing Experiments and Observations.

1. That *animal* and *farinaceous* aliments are more easy of digestion than *vegetable.*

2. That the susceptibility of digestion does not, however, depend altogether upon *natural* or *chemical* distinctions.

3. That digestion is facilitated by *minuteness* of *division* and *tenderness* of *fibre*, and retarded by opposite qualities.

4. That the *ultimate principles* of aliment are always the same, from whatever food they may be obtained.

5. That the action of the stomach, and its fluids are the same on *all kinds* of diet.

6. That the *digestibility* of aliment does not depend upon the *quantity* of nutrient principles that it contains.

7. That the *quantity* of food generally taken, is more than the wants of the system require; and that such excess, if persevered in, generally produces, not only functional aberration, but disease of the coats of the stomach.

8. That *bulk*, as well as *nutriment*, is necessary to the articles of diet.

9. That *oily* food is difficult of digestion, though it contains a large proportion of the nutrient principles.

10. That the *time* required for the digestion of food, is various, depending upon the quantity and quality of the food, state of the stomach, &c.; but that the time ordinarily required for the disposal of a moderate meal of the fibrous parts of meat, with bread, &c., is from three to three and a half hours.

11. That *solid* food, of a certain texture, is easier of digestion, than *fluid.*

12. That stimulating *condiments* are injurious to the healthy stomach.

13. That the use of *ardent spirits always* produces disease of the stomach, if persevered in.

14. That *hunger* is the effect of *distention* of the vessels that secrete the gastric juice.

15. That the processes of *mastication*, *insalivation* and *deglutition*, in an abstract point of view, do not, in any way, affect the digestion of food; or, in other words, when food is introduced directly into the stomach, in a finely divided state, without these previous steps, it is as readily and as perfectly digested as when they have been taken.

16. That *saliva* does not possess the properties of an alimentary solvent.

17. That the *first* stage of digestion is effected in the stomach.

18. That the natural *temperature* of the stomach is 100° Fahrenheit.

19. That tbe temperature is *not elevated* by the ingestion of food.

20. That *exercise elevates* the temperature; and that *sleep* or *rest*, in a recumbent position, *depresses* it.

21. That the *agent* of chymification is the *Gastric Juice.*

22. That it acts as a *solvent* of food, and alters its properties.

23. That its action is facilitated by the *warmth* and *motions* of the stomach.

24. That it contains free *Muriatic Acid* and some other active *chemical* principles.

25. That it is never found *free* in the gastric cavity; but is always excited to discharge itself by the introduction of *food*, or other irritants.

26. That it is secreted from vessels distinct from the mucous follicles.

27. That it is seldom obtained pure, but is generrally mixed with mucus, and sometimes with saliva

When pure, it is capable of being kept for months, and perhaps for years.*

28. That it *coagulates* albumen, and afterwards *dissolves* the *coagulæ.*

29. That it *checks* the progress of putrefaction.

30. That the pure gastric juice is fluid, *clear* and *transparent* ; without *odour* ; a little *salt*, and perceptibly *acid.*

31. That like other chemical agents, it *commences* its action on food, as soon as it comes in *contact* with it.

32. That it is capable of *combining* with a certain and fixed *quantity* of food, and when more aliment is presented for its action than it will dissolve, disturbance of the stomach, or "indigestion," will ensue.

33. That it becomes intimately *mixed* and *blended* with the ingestæ in the stomach, by the motions of that organ.

34. That it is *invariably* the *same substance*, modified only by *admixture* with other fluids.

35. That *gentle exercise* facilitates the digestion of food.

36. That *bile* is not ordinarily found *in the stomach*, and is *not* commonly *necessary* for the digestion of food; but

37. That, when *oily* food has been used, it assists its digestion.

38. That *chyme* is *homogeneous*, but variable in its *colour* and *consistence.*

39. That towards the *latter stages* of chymification, it becomes more *acid* and *stimulating*, and passes more rapidly from the stomach.

40. That *water*, *ardent spirits*, and most other *fluids* are not affected by the gastric juice, but pass from the stomach soon after they have been received.

* I have now (Nov. 1, 1833) in my possession, some clear gastric juice, possessing all its original properties, unchanged and undiminished, which was taken from the stomach in Dec. 1832, about eleven months ago, and has been kept tightly corked in vials.

41. That the *inner coat* of the stomach, is of a pale *pink* colour, varying in its hues, according to its full or empty state.

42. That, in health, it is constantly sheathed with a *mucous* coat.

43. That the gastric juice and mucus are *dissimilar* in their *physical* and *chemical* properties.

44. That the appearance of the interior of the stomach, *in disease*, is essentially different from that of its *healthy* state.

45. That the motions of the stomach produce a constant *churning* of its contents, and *admixture* of food and gastric juice.

46. That these motions are in two directions; *transversely* and *longitudinally*.

47. That the *expulsion* of the chyme is assisted by a *transverse band*, &c.

48. That *chyle* is formed in the duodenum and small intestines, by the action of *bile* and *pancreatic juice*, on the chyme.

49. That crude *chyle* is a *semi-transparent*, *whey coloured* fluid.

50. That it is further changed by the action of the *lacteals* and *mesenteric glands*. This is only an *inference* from the other facts. It has not been the subject of experiment.

51. That *no other* fluid produces the same effect on food that gastric juice does; and that it is the *only solvent* of *aliment*.

I regret, exceedingly, that I have not been able to obtain returns from Professor Berzelius, to whom I transmitted, about seven months ago, a bottle of gastric juice for chemical examination. I could not, however, consistently with the expectations and wishes of my friends, further delay the publication of these experiments.

ERRATA.

Page 5—last line, for Spallanz*i*ni, read Spallanz*a*ni.
67—9th line from bottom, for *u*limentaria, read *a*limentaria.
97—1st line, before *more* read *and.*
104—10th line from bottom, for *was* read *were.*
109—14th " " for *clear* read *clean.*
113—1st line, before *stomach*, read *into the.*
117—1st line, for *Chymification* read *Chylification.*
132—1st line, for 94° read 98°.
" —3d line, Ex. 2d, for 96° read 98°.
175—Ex. 13, 1st line. after *Dec.* read 14.
201—15th line from top, for *half* read *quarter.*
206—Ex. 63, for *Jan.* 9, read *Jan.* 19.
222—17th line from top for *expulsion* read *expulsive,*
228—2d par. 4th line, before *alkaline* read *acid and.*
249—Ex. 39, 1st line, after 2 *o'clock*, read 30 *mins.*
260—18th line from top, for *clear* read *clean.*
262—4th line from bottom, for *No.* 2, read *No.* 1.
264—7th " " for *above* read *about.*
270—15th article of diet, (mean time of stomachic digestion) for 3, 36, read 2, 45.
" —16th " " for 3, 10, read 3, 30.
268—Ex. 60, last line but one, for *on* read *in.*

CONTENTS.

Catalog of DOVER BOOKS

BOOKS EXPLAINING SCIENCE

(Note: The books listed under this category are general introductions, surveys, reviews, and non-technical expositions of science for the interested layman or scientist who wishes to brush up. Dover also publishes the largest list of inexpensive reprints of books on intermediate and higher mathematics, mathematical physics, engineering, chemistry, astronomy, etc., for the professional mathematician or scientist. For our complete Science Catalog, write Dept. catrr., Dover Publications, Inc., 180 Varick Street, New York 14, N. Y.)

CONCERNING THE NATURE OF THINGS, Sir William Bragg. Royal Institute Christmas Lectures by Nobel Laureate. Excellent plain-language introduction to gases, molecules, crystal structure, etc. explains "building blocks" of universe, basic properties of matter, with simplest, clearest examples, demonstrations. 32pp. of photos; 57 figures. 244pp. 5⅜ x 8.
T31 Paperbound **$1.35**

MATTER AND LIGHT, THE NEW PHYSICS, Louis de Broglie. Non-technical explanations by a Nobel Laureate of electro-magnetic theory, relativity, wave mechanics, quantum physics, philosophies of science, etc. Simple, yet accurate introduction to work of Planck, Bohr, Einstein, other modern physicists. Only 2 of 12 chapters require mathematics. 300pp. 5⅜ x 8.
T35 Paperbound **$1.60**

THE COMMON SENSE OF THE EXACT SCIENCES, W. K. Clifford. For 70 years, Clifford's work has been acclaimed as one of the clearest, yet most precise introductions to mathematical symbolism, measurement, surface boundaries, position, space, motion, mass and force, etc. Prefaces by Bertrand Russell and Karl Pearson. Introduction by James Newman. 130 figures. 249pp. 5⅜ x 8.
T61 Paperbound **$1.60**

THE NATURE OF LIGHT AND COLOUR IN THE OPEN AIR, M. Minnaert. What causes mirages? haloes? "multiple" suns and moons? Professor Minnaert explains these and hundreds of other fascinating natural optical phenomena in simple terms, tells how to observe them, suggests hundreds of experiments. 200 illus; 42 photos. xvi + 362pp.
T196 Paperbound **$1.95**

SPINNING TOPS AND GYROSCOPIC MOTION, John Perry. Classic elementary text on dynamics of rotation treats gyroscopes, tops, how quasi-rigidity is induced in paper disks, smoke rings, chains, etc, by rapid motion, precession, earth's motion, etc. Contains many easy-to-perform experiments. Appendix on practical uses of gyroscopes. 62 figures. 128pp.
T416 Paperbound **$1.00**

A CONCISE HISTORY OF MATHEMATICS, D. Struik. This lucid, easily followed history of mathematics from the Ancient Near East to modern times requires no mathematical background itself, yet introduces both mathematicians and laymen to basic concepts and discoveries and the men who made them. Contains a collection of 31 portraits of eminent mathematicians. Bibliography. xix + 299pp. 5⅜ x 8.
T255 Paperbound **$1.75**

THE RESTLESS UNIVERSE, Max Born. A remarkably clear, thorough exposition of gases, electrons, ions, waves and particles, electronic structure of the atom, nuclear physics, written for the layman by a Nobel Laureate. "Much more thorough and deep than most attempts . . . easy and delightful," CHEMICAL AND ENGINEERING NEWS. Includes 7 animated sequences showing motion of molecules, alpha particles, etc. 11 full-page plates of photographs. Total of nearly 600 illus. 315pp. 6⅛ x 9¼.
T412 Paperbound **$2.00**

WHAT IS SCIENCE?, N. Campbell. The role of experiment, the function of mathematics, the nature of scientific laws, the limitations of science, and many other provocative topics are explored without technicalities by an eminent scientist. "Still an excellent introduction to scientific philosophy," H. Margenau in PHYSICS TODAY. 192pp. 5⅜ x 8.
S43 Paperbound **$1.25**

TRADE

FADS AND FALLACIES IN THE NAME OF SCIENCE, Martin Gardner. The standard account of the various cults, quack systems and delusions which have recently masqueraded as science: hollow earth theory, Atlantis, dianetics, Reich's orgone theory, flying saucers, Bridey Murphy, psionics, irridiagnosis, many other fascinating fallacies that deluded tens of thousands. "Should be read by everyone, scientist and non-scientist alike," R. T. Birge, Prof. Emeritus, Univ. of California; Former President, American Physical Society. Formerly titled, "In the Name of Science." Revised and enlarged edition. x + 365pp. 5⅜ x 8.
T394 Paperbound **$1.50**

THE STUDY OF THE HISTORY OF MATHEMATICS, THE STUDY OF THE HISTORY OF SCIENCE, G. Sarton. Two books bound as one. Both volumes are standard introductions to their fields by an eminent science historian. They discuss problems of historical research, teaching, pitfalls, other matters of interest to the historically oriented writer, teacher, or student. Both have extensive bibliographies. 10 illustrations. 188pp. 5⅜ x 8. T240 Paperbound **$1.25**

THE PRINCIPLES OF SCIENCE, W. S. Jevons. Unabridged reprinting of a milestone in the development of symbolic logic and other subjects concerning scientific methodology, probability, inferential validity, etc. Also describes Jevons' "logic machine," an early precursor of modern electronic calculators. Preface by E. Nagel. 839pp. 5⅜ x 8. S446 Paperbound **$2.98**

SCIENCE THEORY AND MAN, Erwin Schroedinger. Complete, unabridged reprinting of "Science and the Human Temperament" plus an additional essay "What is an Elementary Particle?" Nobel Laureate Schroedinger discusses many aspects of modern physics from novel points of view which provide unusual insights for both laymen and physicists. 192 pp. 5⅜ x 8.
T428 Paperbound **$1.35**

BRIDGES AND THEIR BUILDERS, D. B. Steinman & S. R. Watson. Information about ancient, medieval, modern bridges; how they were built; who built them; the structural principles employed; the materials they are built of; etc. Written by one of the world's leading authorities on bridge design and construction. New, revised, expanded edition. 23 photos; 26 line drawings, xvii + 401pp. 5⅜ x 8. T431 Paperbound **$1.95**

HISTORY OF MATHEMATICS, D. E. Smith. Most comprehensive non-technical history of math in English. In two volumes. Vol. I: A chronological examination of the growth of mathematics from primitive concepts up to 1900. Vol. II: The development of ideas in specific fields and areas, up through elementary calculus. The lives and works of over a thousand mathematicians are covered; thousands of specific historical problems and their solutions are clearly explained. Total of 510 illustrations, 1355pp. 5⅜ x 8. Set boxed in attractive container.
T429, T430 Paperbound, the set **$5.00**

PHILOSOPHY AND THE PHYSICISTS, L. S. Stebbing. A philosopher examines the philosophical implications of modern science by posing a lively critical attack on the popular science expositions of Sir James Jeans and Arthur Eddington. xvi + 295pp. 5⅜ x 8.
T480 Paperbound **$1.65**

ON MATHEMATICS AND MATHEMATICIANS, R. E. Moritz. The first collection of quotations by and about mathematicians in English. 1140 anecdotes, aphorisms, definitions, speculations, etc. give both mathematicians and layman stimulating new insights into what mathematics is, and into the personalities of the great mathematicians from Archimedes to Euler, Gauss, Klein, Weierstrass. Invaluable to teachers, writers. Extensive cross index. 410pp. 5⅜ x 8.
T489 Paperbound **$1.95**

NATURAL SCIENCE, BIOLOGY, GEOLOGY, TRAVEL

A SHORT HISTORY OF ANATOMY AND PHYSIOLOGY FROM THE GREEKS TO HARVEY, C. Singer. A great medical historian's fascinating intermediate account of the slow advance of anatomical and physiological knowledge from pre-scientific times to Vesalius, Harvey. 139 unusually interesting illustrations. 221pp. 5⅜ x 8. T389 Paperbound **$1.75**

THE BEHAVIOUR AND SOCIAL LIFE OF HONEYBEES, Ronald Ribbands. The most comprehensive, lucid and authoritative book on bee habits, communication, duties, cell life, motivations, etc. "A MUST for every scientist, experimenter, and educator, and a happy and valuable selection for all interested in the honeybee," AMERICAN BEE JOURNAL. 690-item bibliography. 127 illus.; 11 photographic plates. 352pp. 5⅜ x 8⅜. S410 Clothbound **$4.50**

TRAVELS OF WILLIAM BARTRAM, edited by **Mark Van Doren.** One of the 18th century's most delightful books, and one of the few first-hand sources of information about American geography, natural history, and anthropology of American Indian tribes of the time. "The mind of a scientist with the soul of a poet," John Livingston Lowes. 13 original illustrations, maps. Introduction by Mark Van Doren. 448pp. 5⅜ x 8. T326 Paperbound **$2.00**

STUDIES ON THE STRUCTURE AND DEVELOPMENT OF VERTEBRATES, Edwin Goodrich. The definitive study of the skeleton, fins and limbs, head region, divisions of the body cavity, vascular, respiratory, excretory systems, etc., of vertebrates from fish to higher mammals, by the greatest comparative anatomist of recent times. "The standard textbook," JOURNAL OF ANATOMY. 754 illus. 69-page biographical study. 1186-item bibliography. 2 vols. Total of 906pp. 5⅜ x 8.
Vol. I: S449 Paperbound **$2.50**
Vol. II: S450 Paperbound **$2.50**

LITERATURE

WORLD DRAMA, B. H. Clark. 46 plays from Ancient Greece, Rome, to India, China, Japan. Plays by Aeschylus, Sophocles, Euripides, Aristophanes, Plautus, Marlowe, Jonson, Farquhar, Goldsmith, Cervantes, Molière, Dumas, Goethe, Schiller, Ibsen, many others. One of the most comprehensive collections of important plays from all literature available in English. Over 1/3 of this material is unavailable in any other current edition. Reading lists. 2 volumes. Total of 1364pp. 5⅜ x 8. Vol. I, T57 Paperbound **$2.00**
Vol. II, T59 Paperbound **$2.00**

MASTERS OF THE DRAMA, John Gassner. The most comprehensive history of the drama in print. Covers more than 800 dramatists and over 2000 plays from the Greeks to modern Western, Near Eastern, Oriental drama. Plot summaries, theatre history, etc. "Best of its kind in English," NEW REPUBLIC. 35 pages of bibliography. 77 photos and drawings. Deluxe edition. xxii + 890pp. 5⅜ x 8. T100 Clothbound **$5.95**

THE DRAMA OF LUIGI PIRANDELLO, D. Vittorini. All 38 of Pirandello's plays (to 1935) summarized and analyzed in terms of symbolic techniques, plot structure, etc. The only authorized work. Foreword by Pirandello. Biography. Bibliography. xiii + 350pp. 5⅜ x 8.
T435 Paperbound **$1.98**

ARISTOTLE'S THEORY OF POETRY AND THE FINE ARTS, S. H. Butcher, ed. The celebrated "Butcher translation" faced page by page with the Greek text; Butcher's 300-page introduction to Greek poetic, dramatic thought. Modern Aristotelian criticism discussed by John Gassner. lxxvi + 421pp. 5⅜ x 8.
T42 Paperbound **$2.00**

EUGENE O'NEILL: THE MAN AND HIS PLAYS, B. H. Clark. The first published source-book on O'Neill's life and work. Analyzes each play from the early THE WEB up to THE ICEMAN COMETH. Supplies much information about environmental and dramatic influences. ix + 182pp. 5⅜ x 8. T379 Paperbound **$1.25**

INTRODUCTION TO ENGLISH LITERATURE, B. Dobrée, ed. Most compendious literary aid in its price range. Extensive, categorized bibliography (with entries up to 1949) of more than 5,000 poets, dramatists, novelists, as well as historians, philosophers, economists, religious writers, travellers, and scientists of literary stature. Information about manuscripts, important biographical data. Critical, historical, background works not simply listed, but evaluated. Each volume also contains a long introduction to the period it covers.

Vol. I: **THE BEGINNINGS OF ENGLISH LITERATURE TO SKELTON, 1509, W. L. Renwick. H. Orton.** 450pp. 5⅛ x 7⅛. T75 Clothbound **$3.50**

Vol. II: **THE ENGLISH RENAISSANCE, 1510-1688, V. de Sola Pinto.** 381pp. 5⅛ x 7⅛.
T76 Clothbound **$3.50**

Vol. III: **THE AUGUSTANS AND ROMANTICS, 1689-1830, H. Dyson, J. Butt.** 320pp. 5⅛ x 7⅛.
T77 Clothbound **$3.50**

Vol. IV: **THE VICTORIANS AND AFTER, 1830-1914, E. Batho, B. Dobrée.** 360pp. 5⅛ x 7⅛.
T78 Clothbound **$3.50**

EPIC AND ROMANCE, W. P. Ker. The standard survey of Medieval epic and romance by a foremost authority on Medieval literature. Covers historical background, plot, literary analysis, significance of Teutonic epics, Icelandic sagas, Beowulf, French chansons de geste, the Niebelungenlied, Arthurian romances, much more. 422pp. 5⅜ x 8. T355 Paperbound **$1.95**

THE HEART OF EMERSON'S JOURNALS, Bliss Perry, ed. Emerson's most intimate thoughts, impressions, records of conversations with Channing, Hawthorne, Thoreau, etc., carefully chosen from the 10 volumes of The Journals. "The essays do not reveal the power of Emerson's mind . . .as do these hasty and informal writings," N. Y. TIMES. Preface by B. Perry. 370pp. 5⅜ x 8. T447 Paperbound **$1.85**

A SOURCE BOOK IN THEATRICAL HISTORY, A. M. Nagler. (Formerly, "Sources of Theatrical History.") Over 300 selected passages by contemporary observers tell about styles of acting, direction, make-up, scene designing, etc., in the theatre's great periods from ancient Greece to the Théâtre Libre. "Indispensable complement to the study of drama," EDUCATIONAL THEATRE JOURNAL. Prof. Nagler, Yale Univ. School of Drama, also supplies notes, references. 85 illustrations. 611pp. 5⅜ x 8. T515 Paperbound **$2.75**

THE ART OF THE STORY-TELLER, M. L. Shedlock. Regarded as the finest, most helpful book on telling stories to children, by a great story-teller. How to catch, hold, recapture attention; how to choose material; many other aspects. Also includes: a 99-page selection of Miss Shedlock's most successful stories; extensive bibliography of other stories. xxi + 320pp. 5⅜ x 8. T245 Clothbound **$3.50**

THE DEVIL'S DICTIONARY, Ambrose Bierce. Over 1000 short, ironic definitions in alphabetical order, by America's greatest satirist in the classical tradition. "Some of the most gorgeous witticisms in the English language," H. L. Mencken. 144pp. 5⅜ x 8. T487 Paperbound **$1.00**

THE POLITICAL THOUGHT OF PLATO AND ARISTOTLE, Ernest Barker. The standard, comprehensive exposition of Greek political thought. Covers every aspect of the "Republic" and the "Politics" as well as minor writings, other philosophers, theorists of the period, and the later history of Greek political thought. Unabridged edition. 584pp. 5⅜ x 8.
T521 Paperbound **$1.85**

PHILOSOPHY

THE GIFT OF LANGUAGE, M. Schlauch. (Formerly, "The Gift of Tongues.") A sound, middle-level treatment of linguistic families, word histories, grammatical processes, semantics, language taboos, word-coining of Joyce, Cummings, Stein, etc. 232 bibliographical notes. 350pp. 5⅜ x 8.
T243 Paperbound **$1.85**

THE PHILOSOPHY OF HEGEL, W. T. Stace. The first work in English to give a complete and connected view of Hegel's entire system. Especially valuable to those who do not have time to study the highly complicated original texts, yet want an accurate presentation by a most reputable scholar of one of the most influential 19th century thinkers. Includes a 14 x 20 fold-out chart of Hegelian system. 536pp. 5⅜ x 8.
T254 Paperbound **$2.00**

ARISTOTLE, A. E. Taylor. A lucid, non-technical account of Aristotle written by a foremost Platonist. Covers life and works; thought on matter, form, causes, logic, God, physics, metaphysics, etc. Bibliography. New index compiled for this edition. 128pp. 5⅜ x 8.
T280 Paperbound **$1.00**

GUIDE TO PHILOSOPHY, C. E. M. Joad. This basic work describes the major philosophic problems and evaluates the answers propounded by great philosophers from the Greeks to Whitehead, Russell. "The finest introduction," BOSTON TRANSCRIPT. Bibliography, 592pp. 5⅜ x 8.
T297 Paperbound **$2.00**

LANGUAGE AND MYTH, E. Cassirer. Cassirer's brilliant demonstration that beneath both language and myth lies an unconscious "grammar" of experience whose categories and canons are not those of logical thought. Introduction and translation by Susanne Langer. Index. x + 103pp. 5⅜ x 8.
T51 Paperbound **$1.25**

SUBSTANCE AND FUNCTION, EINSTEIN'S THEORY OF RELATIVITY, E. Cassirer. This double volume contains the German philosopher's profound philosophical formulation of the differences between traditional logic and the new logic of science. Number, space, energy, relativity, many other topics are treated in detail. Authorized translation by W. C. and M. C. Swabey. xii + 465pp. 5⅜ x 8.
T50 Paperbound **$2.00**

THE PHILOSOPHICAL WORKS OF DESCARTES. The definitive English edition, in two volumes, of all major philosophical works and letters of René Descartes, father of modern philosophy of knowledge and science. Translated by E. S. Haldane and G. Ross. Introductory notes. Total of 842pp. 5⅜ x 8.
T71 Vol. 1, Paperbound **$2.00**
T72 Vol. 2, Paperbound **$2.00**

ESSAYS IN EXPERIMENTAL LOGIC, J. Dewey. Based upon Dewey's theory that knowledge implies a judgment which in turn implies an inquiry, these papers consider such topics as the thought of Bertrand Russell, pragmatism, the logic of values, antecedents of thought, data and meanings. 452pp. 5⅜ x 8.
T73 Paperbound **$1.95**

THE PHILOSOPHY OF HISTORY, G. W. F. Hegel. This classic of Western thought is Hegel's detailed formulation of the thesis that history is not chance but a rational process, the realization of the Spirit of Freedom. Translated and introduced by J. Sibree. Introduction by C. Hegel. Special introduction for this edition by Prof. Carl Friedrich, Harvard University. xxxix + 447pp. 5⅜ x 8.
T112 Paperbound **$1.85**

THE WILL TO BELIEVE and HUMAN IMMORTALITY, W. James. Two of James's most profound investigations of human belief in God and immortality, bound as one volume. Both are powerful expressions of James's views on chance vs. determinism, pluralism vs. monism, will and intellect, arguments for survival after death, etc. Two prefaces. 429pp. 5⅜ x 8.
T294 Clothbound **$3.75**
T291 Paperbound **$1.65**

INTRODUCTION TO SYMBOLIC LOGIC, S. Langer. A lucid, general introduction to modern logic, covering forms, classes, the use of symbols, the calculus of propositions, the Boole-Schroeder and the Russell-Whitehead systems, etc. "One of the clearest and simplest introductions," MATHEMATICS GAZETTE. Second, enlarged, revised edition. 368pp. 5⅜ x 8.
S164 Paperbound **$1.75**

MIND AND THE WORLD-ORDER, C. I. Lewis. Building upon the work of Peirce, James, and Dewey, Professor Lewis outlines a theory of knowledge in terms of "conceptual pragmatism," and demonstrates why the traditional understanding of the a priori must be abandoned. Appendices. xiv + 446pp. 5⅜ x 8.
T359 Paperbound **$1.95**

THE GUIDE FOR THE PERPLEXED, M. Maimonides One of the great philosophical works of all time, Maimonides' formulation of the meeting-ground between Old Testament and Aristotelian thought is essential to anyone interested in Jewish, Christian, and Moslem thought in the Middle Ages. 2nd revised edition of the Friedländer translation. Extensive introduction. lix + 414pp. 5⅜ x 8.
T351 Paperbound **$1.85**

ORIENTALIA AND RELIGION

THE MYSTERIES OF MITHRA, F. Cumont. The great Belgian scholar's definitive study of the Persian mystery religion that almost vanquished Christianity in the ideological struggle for the Roman Empire. A masterpiece of scholarly detection that reconstructs secret doctrines, organization, rites. Mithraic art is discussed and analyzed. 70 illus. 239pp. 5⅜ x 8.
T323 Paperbound **$1.85**

CHRISTIAN AND ORIENTAL PHILOSOPHY OF ART. A. K. Coomaraswamy. The late art historian and orientalist discusses artistic symbolism, the role of traditional culture in enriching art, medieval art, folklore, philosophy of art, other similar topics. Bibliography. 148pp. 5⅜ x 8.
T378 Paperbound **$1.25**

TRANSFORMATION OF NATURE IN ART, A. K. Coomaraswamy. A basic work on Asiatic religious art. Includes discussions of religious art in Asia and Medieval Europe (exemplified by Meister Eckhart), the origin and use of images in Indian art, Indian Medieval aesthetic manuals, and other fascinating, little known topics. Glossaries of Sanskrit and Chinese terms. Bibliography. 41pp. of notes. 245pp. 5⅜ x 8. T368 Paperbound **$1.75**

ORIENTAL RELIGIONS IN ROMAN PAGANISM, F. Cumont. This well-known study treats the ecstatic cults of Syria and Phrygia (Cybele, Attis, Adonis, their orgies and mutilatory rites); the mysteries of Egypt (Serapis, Isis, Osiris); Persian dualism; Mithraic cults; Hermes Trismegistus, Ishtar, Astarte, etc. and their influence on the religious thought of the Roman Empire. Introduction. 55pp. of notes; extensive bibliography. xxiv + 298pp. 5⅜ x 8.
T321 Paperbound **$1.75**

ANTHROPOLOGY, SOCIOLOGY, AND PSYCHOLOGY

PRIMITIVE MAN AS PHILOSOPHER, P. Radin. A standard anthropological work based on Radin's investigations of the Winnebago, Maori, Batak, Zuni, other primitive tribes. Describes primitive thought on the purpose of life, marital relations, death, personality, gods, etc. Extensive selections of original primitive documents. Bibliography. xviii + 420pp. 5⅜ x 8.
T392 Paperbound **$2.00**

PRIMITIVE RELIGION, P. Radin. Radin's thoroughgoing treatment of supernatural beliefs, shamanism, initiations, religious expression, etc. in primitive societies. Arunta, Ashanti, Aztec, Bushman, Crow, Fijian, many other tribes examined. "Excellent," NATURE. New preface by the author. Bibliographic notes. x + 322pp. 5⅜ x 8. T393 Paperbound **$1.85**

SEX IN PSYCHO-ANALYSIS, S. Ferenczi. (Formerly, "Contributions to Psycho-analysis.") 14 selected papers on impotence, transference, analysis and children, dreams, obscene words, homosexuality, paranoia, etc. by an associate of Freud. Also included: THE DEVELOPMENT OF PSYCHO-ANALYSIS, by Ferenczi and Otto Rank. Two books bound as one. Total of 406pp. 5⅜ x 8. T324 Paperbound **$1.85**

THE PRINCIPLES OF PSYCHOLOGY, William James. The complete text of the famous "long course," one of the great books of Western thought. An almost incredible amount of information about psychological processes, the stream of consciousness, habit, time perception, memory, emotions, reason, consciousness of self, abnormal phenomena, and similar topics. Based on James's own discoveries integrated with the work of Descartes, Locke, Hume, Royce, Wundt, Berkeley, Lotse, Herbart, scores of others. "A classic of interpretation," PSYCHIATRIC QUARTERLY. 94 illus. 1408pp. 2 volumes. 5⅜ x 8.
T381 Vol. 1, Paperbound **$2.50**
T382 Vol. 2, Paperbound **$2.50**

THE POLISH PEASANT IN EUROPE AND AMERICA, W. I. Thomas, F. Znaniecki. Monumental sociological study of peasant primary groups (family and community) and the disruptions produced by a new industrial system and emigration to America, by two of the foremost sociologists of recent times. One of the most important works in sociological thought. Includes hundreds of pages of primary documentation; point by point analysis of causes of social decay, breakdown of morality, crime, drunkenness, prostitution, etc. 2nd revised edition. 2 volumes. Total of 2250pp. 6 x 9. T478 2 volume set, Clothbound **$12.50**

FOLKWAYS, W. G. Sumner. The great Yale sociologist's detailed exposition of thousands of social, sexual, and religious customs in hundreds of cultures from ancient Greece to Modern Western societies. Preface by A. G. Keller. Introduction by William Lyon Phelps. 705pp. 5⅜ x 8. S508 Paperbound **$2.49**

BEYOND PSYCHOLOGY, Otto Rank. The author, an early associate of Freud, uses psychoanalytic techniques of myth-analysis to explore ultimates of human existence. Treats love, immortality, the soul, sexual identity, kingship, sources of state power, many other topics which illuminate the irrational basis of human existence. 291pp. 5⅜ x 8. T485 Paperbound **$1.75**

ILLUSIONS AND DELUSIONS OF THE SUPERNATURAL AND THE OCCULT, D. H. Rawcliffe. A rational, scientific examination of crystal gazing, automatic writing, table turning, stigmata, the Indian rope trick, dowsing, telepathy, clairvoyance, ghosts, ESP, PK, thousands of other supposedly occult phenomena. Originally titled "The Psychology of the Occult." 14 illustrations. 551pp. 5⅜ x 8. T503 Paperbound **$2.00**

YOGA: A SCIENTIFIC EVALUATION, Kovoor T. Behanan. A scientific study of the physiological and psychological effects of Yoga discipline, written under the auspices of the Yale University Institute of Human Relations. Foreword by W. A. Miles, Yale Univ. 17 photographs. 290pp. 5⅜ x 8. T505 Paperbound **$1.65**

HOAXES, C. D. MacDougall. Delightful, entertaining, yet scholarly exposition of how hoaxes start, why they succeed, documented with stories of hundreds of the most famous hoaxes. "A stupendous collection . . . and shrewd analysis, "NEW YORKER. New, revised edition. 54 photographs. 320pp. 5⅜ x 8. T465 Paperbound **$1.75**

CREATIVE POWER: THE EDUCATION OF YOUTH IN THE CREATIVE ARTS, Hughes Mearns. Named by the National Education Association as one of the 20 foremost books on education in recent times. Tells how to help children express themselves in drama, poetry, music, art, develop latent creative power. Should be read by every parent, teacher. New, enlarged, revised edition. Introduction. 272pp. 5⅜ x 8. T490 Paperbound **$1.50**

LANGUAGES

NEW RUSSIAN-ENGLISH, ENGLISH-RUSSIAN DICTIONARY, M. A. O'Brien. Over 70,000 entries in new orthography! Idiomatic usages, colloquialisms. One of the few dictionaries that indicate accent changes in conjugation and declension. "One of the best," Prof. E. J. Simmons, Cornell. First names, geographical terms, bibliography, many other features. 738pp. 4½ x 6¼. T208 Paperbound **$2.00**

MONEY CONVERTER AND TIPPING GUIDE FOR EUROPEAN TRAVEL, C. Vomacka. Invaluable, handy source of currency regulations, conversion tables, tipping rules, postal rates, much other travel information for every European country plus Israel, Egypt and Turkey. 128pp. 3½ x 5¼. T260 Paperbound **60¢**

MONEY CONVERTER AND TIPPING GUIDE FOR TRAVEL IN THE AMERICAS (including the United States and Canada), **C. Vomacka.** The information you need for informed and confident travel in the Americas: money conversion tables, tipping guide, postal, telephone rates, etc. 128pp. 3½ x 5¼. T261 Paperbound **65¢**

DUTCH-ENGLISH, ENGLISH-DUTCH DICTIONARY, F. G. Renier. The most convenient, practical Dutch-English dictionary on the market. New orthography. More than 60,000 entries: idioms, compounds, technical terms, etc. Gender of nouns indicated. xviii + 571pp. 5½ x 6¼. T224 Clothbound **$2.50**

LEARN DUTCH!, F. G. Renier. The most satisfactory and easily-used grammar of modern Dutch. Used and recommended by the Fulbright Committee in the Netherlands. Over 1200 simple exercises lead to mastery of spoken and written Dutch. Dutch-English, English-Dutch vocabularies. 181pp. 4¼ x 7¼. T441 Clothbound **$1.75**

PHRASE AND SENTENCE DICTIONARY OF SPOKEN RUSSIAN, English-Russian, Russian-English. Based on phrases and complete sentences, rather than isolated words; recognized as one of the best methods of learning the idiomatic speech of a country. Over 11,500 entries, indexed by single words, with more than 32,000 English and Russian sentences and phrases, in immediately usable form. Probably the largest list ever published. Shows accent changes in conjugation and declension; irregular forms listed in both alphabetical place and under main form of word. 15,000 word introduction covering Russian sounds, writing, grammar, syntax. 15-page appendix of geographical names, money, important signs, given names, foods, special Soviet terms, etc. Travellers, businessmen, students, government employees have found this their best source for Russian expressions. Originally published as U.S. Government Technical Manual TM 30-944. iv + 573pp. 5⅝ x 8⅜. T496 Paperbound **$2.75**

PHRASE AND SENTENCE DICTIONARY OF SPOKEN SPANISH, Spanish-English, English-Spanish. Compiled from spoken Spanish, emphasizing idiom and colloquial usage in both Castilian and Latin-American. More than 16,000 entries containing over 25,000 idioms—the largest list of idiomatic constructions ever published. Complete sentences given, indexed under single words —language in immediately usable form, for travellers, businessmen, students, etc. 25-page introduction provides rapid survey of sounds, grammar, syntax, with full consideration of irregular verbs. Especially apt in modern treatment of phrases and structure. 17-page glossary gives translations of geographical names, money values, numbers, national holidays, important street signs, useful expressions of high frequency, plus unique 7-page glossary of Spanish and Spanish-American foods and dishes. Originally published as U.S. Government Technical Manual TM 30-900. iv + 513pp. 5⅝ x 8⅜. T495 Paperbound **$1.75**

CATALOG OF

SAY IT language phrase books

"SAY IT" in the foreign language of your choice! We have sold over ½ million copies of these popular, useful language books. They will not make you an expert linguist overnight, but they do cover most practical matters of everyday life abroad.

Over 1000 useful phrases, expressions, with additional variants, substitutions.

Modern! Useful! Hundreds of phrases not available in other texts: "Nylon," "air-conditioned," etc.

The ONLY inexpensive phrase book **completely indexed.** Everything is available at a flip of your finger, ready for use.

Prepared by native linguists, travel experts.

Based on years of travel experience abroad.

This handy phrase book may be used by itself, or it may supplement any other text or course; it provides a living element. Used by many colleges and institutions: Hunter College; Barnard College; Army Ordnance School, Aberdeen; and many others.

Available, 1 book per language:

Danish (T818) 75¢
Dutch T(817) 75¢
English (for German-speaking people) (T801) 60¢
English (for Italian-speaking people) (T816) 60¢
English (for Spanish-speaking people) (T802) 60¢
Esperanto (T820) 75¢
French (T803) 60¢
German (T804) 60¢
Modern Greek (T813) 75¢
Hebrew (T805) 60¢
Italian (T806) 60¢
Japanese (T807) 60¢
Norwegian (T814) 75¢
Russian (T810) 75¢
Spanish (T811) 60¢
Turkish (T821) 75¢
Yiddish (T815) 75¢
Swedish (T812) 75¢
Polish (T808) 75¢
Portuguese (T809) 75¢

LISTEN & LEARN language record sets

LISTEN & LEARN is the only language record course designed especially to meet your travel needs, or help you learn essential foreign language quickly by yourself, or in conjunction with any school course, by means of the automatic association method. Each set contains three 33⅓ rpm long- playing records — 1½ hours of recorded speech by eminent native speakers who are professors at Columbia, N.Y.U., Queens College and other leading universities. The sets are priced far below other sets of similar quality, yet they contain many special features not found in other record sets:

* Over 800 selected phrases and sentences, a basic vocabulary of over 3200 words.
* Both English and foreign language recorded; with a pause for your repetition.
* Designed for persons with limited time; no time wasted on material you cannot use immediately.
* Living, modern expressions that answer modern needs: drugstore items, "air-conditioned," etc.
* 128-196 page manuals contain everything on the records, plus simple pronunciation guides.
* Manual is fully indexed; find the phrase you want instantly.
* High fidelity recording—equal to any records costing up to $6 each.

The phrases on these records cover 41 different categories useful to the traveller or student interested in learning the living, spoken language: greetings, introductions, making yourself understood, passing customs, planes, trains, boats, buses, taxis, nightclubs, restaurants, menu items, sports, concerts, cameras, automobile travel, repairs, drugstores, doctors, dentists, medicines, barber shops, beauty parlors, laundries, many, many more.

"Excellent . . . among the very best on the market," Prof. Mario Pei, Dept. of Romance Languages, Columbia University. "Inexpensive and well-done . . . an ideal present," CHICAGO SUNDAY TRIBUNE. "More genuinely helpful than anything of its kind which I have previously encountered," Sidney Clark, well-known author of "ALL THE BEST" travel books. Each set contains 3 33⅓ rpm pure vinyl records, 128- 196 page with full record text, and album. One language per set. LISTEN & LEARN record sets are now available in—

FRENCH	the set $4.95	**GERMAN**	the set $4.95
ITALIAN	the set $4.95	**SPANISH**	the set $4.95
RUSSIAN	the set $5.95	**JAPANESE** *	the set $5.95

* Available Sept. 1, 1959

UNCONDITIONAL GUARANTEE: Dover Publications stands behind every Listen and Learn record set. If you are dissatisfied with these sets for any reason whatever, return them within 10 days and your money will be refunded in full.

ART HISTORY

STICKS AND STONES, Lewis Mumford. An examination of forces influencing American architecture: the medieval tradition in early New England, the classical influence in Jefferson's time, the Brown Decades, the imperial facade, the machine age, etc. "A truly remarkable book," SAT. REV. OF LITERATURE. 2nd revised edition. 21 illus. xvii + 228pp. 5⅜ x 8.
T202 Paperbound **$1.60**

THE AUTOBIOGRAPHY OF AN IDEA, Louis Sullivan. The architect whom Frank Lloyd Wright called "the master," records the development of the theories that revolutionized America's skyline. 34 full-page plates of Sullivan's finest work. New introduction by R. M. Line. xiv + 335pp. 5⅜ x 8.
T281 Paperbound **$1.85**

THE MATERIALS AND TECHNIQUES OF MEDIEVAL PAINTING, D. V. Thompson. An invaluable study of carriers and grounds, binding media, pigments, metals used in painting, al fresco and al secco techniques, burnishing, etc. used by the medieval masters. Preface by Bernard Berenson. 239pp. 5⅜ x 8.
T327 Paperbound **$1.85**

PRINCIPLES OF ART HISTORY, H. Wölfflin. This remarkably instructive work demonstrates the tremendous change in artistic conception from the 14th to the 18th centuries, by analyzing 164 works by Botticelli, Dürer, Hobbema, Holbein, Hals, Titian, Rembrandt, Vermeer, etc., and pointing out exactly what is meant by "baroque," "classic," "primitive," "picturesque," and other basic terms of art history and criticism. "A remarkable lesson in the art of seeing," SAT. REV. OF LITERATURE. Translated from the 7th German edition. 150 illus. 254pp. 6⅛ x 9¼.
T276 Paperbound **$2.00**

FOUNDATIONS OF MODERN ART, A. Ozenfant. Stimulating discussion of human creativity from paleolithic cave painting to modern painting, architecture, decorative arts. Fully illustrated with works of Gris, Lipchitz, Leger, Picasso, primitive, modern artifacts, architecture, industrial art, much more. 226 illustrations. 368pp. 6⅛ x 9¼.
T215 Paperbound **$1.95**

HANDICRAFTS, APPLIED ART, ART SOURCES, ETC.

WILD FOWL DECOYS, J. Barber. The standard work on this fascinating branch of folk art, ranging from Indian mud and grass devices to realistic wooden decoys. Discusses styles, types, periods; gives full information on how to make decoys. 140 illustrations (including 14 new plates) show decoys and provide full sets of plans for handicrafters, artists, hunters, and students of folk art. 281pp. 7⅞ x 10¾. Deluxe edition.
T11 Clothbound **$8.50**

METALWORK AND ENAMELLING, H. Maryon. Probably the best book ever written on the subject. Tells everything necessary for the home manufacture of jewelry, rings, ear pendants, bowls, etc. Covers materials, tools, soldering, filigree, setting stones, raising patterns, repoussé work, damascening, niello, cloisonné, polishing, assaying, casting, and dozens of other techniques. The best substitute for apprenticeship to a master metalworker. 363 photos and figures. 374pp. 5½ x 8½.
T183 Clothbound **$7.50**

SHAKER FURNITURE, E. D. and F. Andrews. The most illuminating study of Shaker furniture ever written. Covers chronology, craftsmanship, houses, shops, etc. Includes over 200 photographs of chairs, tables, clocks, beds, benches, etc. "Mr. & Mrs. Andrews know all there is to know about Shaker furniture," Mark Van Doren, NATION. 48 full-page plates. 192pp. Deluxe cloth binding. 7⅞ x 10¾.
T7 Clothbound **$6.00**

PRIMITIVE ART, Franz Boas. A great American anthropologist covers theory, technical virtuosity, styles, symbolism, patterns, etc. of primitive art. The more than 900 illustrations will interest artists, designers, craftworkers. Over 900 illustrations. 376pp. 5⅜ x 8.
T25 Paperbound **$1.95**

ON THE LAWS OF JAPANESE PAINTING, H. Bowie. The best possible substitute for lessons from an oriental master. Treats both spirit and technique; exercises for control of the brush; inks, brushes, colors; use of dots, lines to express whole moods, etc. 220 illus. 132pp. 6⅛ x 9¼.
T30 Paperbound **$1.95**

HANDBOOK OF ORNAMENT, F. S. Meyer. One of the largest collections of copyright-free traditional art: over 3300 line cuts of Greek, Roman, Medieval, Renaissance, Baroque, 18th and 19th century art motifs (tracery, geometric elements, flower and animal motifs, etc.) and decorated objects (chairs, thrones, weapons, vases, jewelry, armor, etc.). Full text. 3300 illustrations. 562pp. 5⅜ x 8.
T302 Paperbound **$2.00**

THREE CLASSICS OF ITALIAN CALLIGRAPHY. Oscar Ogg, ed. Exact reproductions of three famous Renaissance calligraphic works: Arrighi's OPERINA and IL MODO, Tagliente's LO PRESENTE LIBRO, and Palatino's LIBRO NUOVO. More than 200 complete alphabets, thousands of lettered specimens, in Papal Chancery and other beautiful, ornate handwriting. Introduction. 245 plates. 282pp. 6⅛ x 9¼.
T212 Paperbound **$1.95**

THE HISTORY AND TECHNIQUES OF LETTERING, A. Nesbitt. A thorough history of lettering from the ancient Egyptians to the present, and a 65-page course in lettering for artists. Every major development in lettering history is illustrated by a complete alphabet. Fully analyzes such masters as Caslon, Koch, Garamont, Jenson, and many more. 89 alphabets, 165 other specimens. 317pp. 5⅜ x 8.
T427 Paperbound **$2.00**

LETTERING AND ALPHABETS, J. A. Cavanagh. An unabridged reissue of "Lettering," containing the full discussion, analysis, illustration of 89 basic hand lettering tyles based on Caslon, Bodoni, Gothic, many other types. Hundreds of technical hints on construction, strokes, pens, brushes, etc. 89 alphabets, 72 lettered specimens, which may be reproduced permission-free. 121pp. 9¾ x 8. T53 Paperbound **$1.25**

THE HUMAN FIGURE IN MOTION, Eadweard Muybridge. The largest collection in print of Muybridge's famous high-speed action photos. 4789 photographs in more than 500 action-strip-sequences (at shutter speeds up to 1/6000th of a second) illustrate men, women, children—mostly undraped—performing such actions as walking, running, getting up, lying down, carrying objects, throwing, etc. "An unparalleled dictionary of action for all artists," AMERICAN ARTIST. 390 full-page plates, with 4789 photographs. Heavy glossy stock, reinforced binding with headbands. 7⅞ x 10¾. T204 Clothbound **$10.00**

ANIMALS IN MOTION, Eadweard Muybridge. The largest collection of animal action photos in print. 34 different animals (horses, mules, oxen, goats, camels, pigs, cats, lions, gnus, deer, monkeys, eagles—and 22 others) in 132 characteristic actions. All 3919 photographs are taken in series at speeds up to 1/1600th of a second, offering artists, biologists, cartoonists a remarkable opportunity to see exactly how an ostrich's head bobs when running, how a lion puts his foot down, how an elephant's knee bends, how a bird flaps his wings, thousands of other hard-to-catch details. "A really marvelous series of plates," NATURE. 380 full-pages of plates. Heavy glossy stock, reinforced binding with headbands. 7⅞ x10¾. T203 Clothbound **$10.00**

THE BOOK OF SIGNS, R. Koch. 493 symbols—crosses, monograms, astrological, biological symbols, runes, etc.—from ancient manuscripts, cathedrals, coins, catacombs, pottery. May be reproduced permission-free. 493 illustrations by Fritz Kredel. 104pp. 6⅛ x 9¼. T162 Paperbound **$1.00**

A HANDBOOK OF EARLY ADVERTISING ART, C. P. Hornung. The largest collection of copyright-free early advertising art ever compiled. Vol. I: 2,000 illustrations of animals, old automobiles, buildings, allegorical figures, fire engines, Indians, ships, trains, more than 33 other categories! Vol II: Over 4,000 typographical specimens; 600 Roman, Gothic, Barnum, Old English faces; 630 ornamental type faces; hundreds of scrolls, initials, flourishes, etc. "A remarkable collection," PRINTERS' INK.

Vol. I: Pictorial Volume. Over 2000 illustrations. 256pp. 9 x 12. T122 Clothbound **$10.00**

Vol. II: Typographical Volume. Over 4000 speciments. 319pp. 9 x 12. T123 Clothbound **$10.00**
Two volume set, Clothbound, only **$18.50**

DESIGN FOR ARTISTS AND CRAFTSMEN, L. Wolchonok. The most thorough course on the creation of art motifs and designs. Shows you step-by-step, with hundreds of examples and 113 detailed exercises, how to create original designs from geometric patterns, plants, birds, animals, humans, and man-made objects. "A great contribution to the field of design and crafts," N. Y. SOCIETY OF CRAFTSMEN. More than 1300 entirely new illustrations. xv + 207pp. 7⅞ x 10¾. T274 Clothbound **$4.95**

HANDBOOK OF DESIGNS AND DEVICES, C. P. Hornung. A remarkable working collection of 1836 basic designs and variations, all copyright-free. Variations of circle, line, cross, diamond, swastika, star, scroll, shield, many more. Notes on symbolism. "A necessity to every designer who would be original without having to labor heavily," ARTIST and ADVERTISER. 204 plates. 240pp. 5⅜ x 8. T125 Paperbound **$1.90**

THE UNIVERSAL PENMAN, George Bickham. Exact reproduction of beautiful 18th century book of handwriting. 22 complete alphabets in finest English roundhand, other scripts, over 2000 elaborate flourishes, 122 calligraphic illustrations, etc. Material is copyright-free. "An essential part of any art library, and a book of permanent value," AMERICAN ARTIST. 212 plates. 224pp. 9 x 13¾. T20 Clothbound **$10.00**

AN ATLAS OF ANATOMY FOR ARTISTS, F. Schider. This standard work contains 189 full-page plates, more than 647 illustrations of all aspects of the human skeleton, musculature, cutaway portions of the body, each part of the anatomy, hand forms, eyelids, breasts, location of muscles under the flesh, etc. 59 plates illustrate how Michelangelo, da Vinci, Goya, 15 others, drew human anatomy. New 3rd edition enlarged by 52 new illustrations by Cloquet, Barcsay. "The standard reference tool," AMERICAN LIBRARY ASSOCIATION. "Excellent," AMERICAN ARTIST. 189 plates, 647 illustrations. xxvi + 192pp. 7⅞ x 10⅝. T241 Clothbound **$6.00**

AN ATLAS OF ANIMAL ANATOMY FOR ARTISTS, W. Ellenberger, H. Baum, H. Dittrich. The largest, richest animal anatomy for artists in English. Form, musculature, tendons, bone structure, expression, detailed cross sections of head, other features, of the horse, lion, dog, cat, deer, seal, kangaroo, cow, bull, goat, monkey, hare, many other animals. "Highly recommended," DESIGN. Second, revised, enlarged edition with new plates from Cuvier, Stubbs, etc. 288 illustrations. 153pp. 11⅜ x 9. T82 Clothbound **$6.00**

ANIMAL DRAWING: ANATOMY AND ACTION FOR ARTISTS, C. R. Knight. 158 studies, with full accompanying text, of such animals as the gorilla, bear, bison, dromedary, camel, vulture, pelican, iguana, shark, etc., by one of the greatest modern masters of animal drawing. Innumerable tips on how to get life expression into your work. "An excellent reference work,' SAN FRANCISCO CHRONICLE. 158 illustrations. 156pp. 10½ x 8½. T426 Paperbound **$2.00**

PUZZLES, GAMES, AND ENTERTAINMENTS

MATHEMATICS, MAGIC AND MYSTERY, Martin Gardner. Astonishing feats of mind reading, mystifying "magic" tricks, are often based on mathematical principles anyone can learn. This book shows you how to perform scores of tricks with cards, dice, coins, knots, numbers, etc., by using simple principles from set theory, theory of numbers, topology, other areas of mathematics, fascinating in themselves. No special knowledge required. 135 illus. 186pp. 5⅜ x 8.
T335 Paperbound **$1.00**

MATHEMATICAL PUZZLES FOR BEGINNERS AND ENTHUSIASTS, G. Mott-Smoth. Test your problem-solving techniques and powers of inference on 188 challenging, amusing puzzles based on algebra, dissection of plane figures, permutations, probabilities, etc. Appendix of primes, square roots, etc. 135 illus. 2nd revised edition. 248pp. 5⅜ x 8.
T198 Paperbound **$1.00**

LEARN CHESS FROM THE MASTERS, F. Reinfeld. Play 10 games against Marshall, Bronstein, Najdorf, other masters, and grade yourself on each move. Detailed annotations reveal principles of play, strategy, etc. as you proceed. An excellent way to get a real insight into the game. Formerly titled, "Chess by Yourself." 91 diagrams. vii + 144pp. 5⅜ x 8.
T362 Paperbound **$1.00**

REINFELD ON THE END GAME IN CHESS, F. Reinfeld. 62 end games of Alekhine, Tarrasch, Morphy, other masters, are carefully analyzed with emphasis on transition from middle game to end play. Tempo moves, queen endings, weak squares, other basic principles clearly illustrated. Excellent for understanding why some moves are weak or incorrect, how to avoid errors. Formerly titled, "Practical End-game Play." 62 diagrams. vi + 177pp. 5⅜ x 8.
T417 Paperbound **$1.25**

101 PUZZLES IN THOUGHT AND LOGIC, C. R. Wylie, Jr. Brand new puzzles you need no special knowledge to solve! Each one is a gem of ingenuity that will really challenge your problem-solving technique. Introduction with simplified explanation of scientic puzzle solving. 128pp. 5⅜ x 8.
T167 Paperbound **$1.00**

THE COMPLETE NONSENSE OF EDWARD LEAR. The only complete edition of this master of gentle madness at a popular price. The Dong with the Luminous Nose, The Jumblies, The Owl and the Pussycat, hundreds of other bits of wonderful nonsense. 214 limericks, 3 sets of Nonsense Botany, 5 Nonsense Alphabets, 546 fantastic drawings, much more. 320pp. 5⅜ x 8.
T167 Paperbound **$1.00**

28 SCIENCE FICTION STORIES OF H. G. WELLS. Two complete novels, "Men Like Gods" and "Star Begotten," plus 26 short stories by the master science-fiction writer of all time. Stories of space, time, future adventure that are among the all-time classics of science fiction. 928pp. 5⅜ x 8.
T265 Clothbound **$3.95**

SEVEN SCIENCE FICTION NOVELS, H. G. Wells. Unabridged texts of "The Time Machine," "The Island of Dr. Moreau," "First Men in the Moon," "The Invisible Man," "The War of the Worlds," "The Food of the Gods," "In the Days of the Comet." "One will have to go far to match this for entertainment, excitement, and sheer pleasure," N. Y. TIMES. 1015pp. 5⅜ x 8.
T264 Clothbound **$3.95**

MATHEMAGIC, MAGIC PUZZLES, AND GAMES WITH NUMBERS, R. V. Heath. More than 60 new puzzles and stunts based on number properties: multiplying large numbers mentally, finding the date of any day in the year, etc. Edited by J. S. Meyer. 76 illus. 129pp. 5⅜ x 8.
T110 Paperbound **$1.00**

FIVE ADVENTURE NOVELS OF H. RIDER HAGGARD. The master story-teller's five best tales of mystery and adventure set against authentic African backgrounds: "She," "King Solomon's Mines," "Allan Quatermain," "Allan's Wife," "Maiwa's Revenge." 821pp. 5⅜ x 8.
T108 Clothbound **$3.95**

WIN AT CHECKERS, M. Hopper. (Formerly "Checkers.") The former World's Unrestricted Checker Champion gives you valuable lessons in openings, traps, end games, ways to draw when you are behind, etc. More than 100 questions and answers anticipate your problems. Appendix. 75 problems diagrammed, solved. 79 figures. xi + 107pp. 5⅜ x 8.
T363 Paperbound **$1.00**

CRYPTOGRAPHY, L. D. Smith. Excellent introductory work on ciphers and their solution, history of secret writing, techniques, etc. Appendices on Japanese methods, the Baconian cipher, frequency tables. Bibliography. Over 150 problems, solutions. 160pp. 5⅜ x 8.
T247 Paperbound **$1.00**

CRYPTANALYSIS, H. F. Gaines. (Formerly, "Elementary Cryptanalysis.") The best book available on cryptograms and how to solve them. Contains all major techniques: substitution, transposition, mixed alphabets, multafid, Kasiski and Vignere methods, etc. Word frequency appendix. 167 problems, solutions. 173 figures. 236pp. 5⅜ x 8.
T97 Paperbound **$1.95**

FLATLAND, E. A. Abbot. The science-fiction classic of life in a 2-dimensional world that is considered a first-rate introduction to relativity and hyperspace, as well as a scathing satire on society, politics and religion. 7th edition. 16 illus. 128pp. 5⅜ x 8.
T1 Paperbound **$1.00**